Praise for

"The author's prose is always crystal-clear and sometimes moving, particularly when he discusses the ways in which his quest revitalized his life in the face of physical decline. An inspiriting story related with journalistic rigor and disarming frankness."

- *Kirkus Reviews*

"Peter Hunt has written a touching, well-crafted book that navigates geographical and human landscapes in his quest to find a lost military aircraft underwater while also dealing with the devastating challenges and uncertainties of his battle with Parkinson's disease."

- Bernie Chowdhury, Author of *The Last Dive*

"Candor combines with a dry sense of humor to create a motivational and inspirational message that causes the reader to think about their own commitments to life while looking forward to every page. *The Lost Intruder* is a lesson for all of us."

- Ken Waidelich, Editor of *The Windscreen,*
Journal of the Intruder Association

"It is the fascinating 'story within the story' that makes this unique tale a must-read and a testament to the capacity of the human spirit. As the pilot of 510 on that fateful day in November 1989, I thought I knew the whole story, but *The Lost Intruder* brings the account to its true conclusion."

- Denby Starling, Vice Admiral (USN, retired)
and former 510 pilot.

"Ejecting from 510 inspired me to take up scuba diving with the unrequited romantic notion that I might one day stumble across my old jet's wreckage. Pete Hunt has turned that dream into reality. From studying the Navy's failed search through the adversity of a debilitating disease, Pete demonstrates that he is a contender in every sense of the word."

- Chris Eagle, Author/Professor of Computer Science at the Naval Post-Graduate School and former 510 bombardier/navigator.

The Lost Intruder

The Lost Intruder

The Search for a Missing Navy Jet

Peter M. Hunt

Cover A-6 photo courtesy of David F. Brown

Copyright 2017 Peter M. Hunt
All rights reserved
ISBN-13: 9781546334972
ISBN-10: 1546334971

Sources, terms, and acknowledgments

General A-6 history and descriptions of aircraft carrier and squadron operations come primarily from personal knowledge and Internet research. Medical explanations of Parkinson's disease and Deep Brain Stimulation surgery come almost entirely from my own experience. Although I am not a medical professional, this is an accurate account of what I believed to be true about Parkinson's disease at the time. The descriptions of technical aspects of the disease and treatment are a layperson's best understanding of how often complex neurologic interactions work.

Certain Navy terms have been renamed for ease in remembering them, such as *Flight Mishap Report* instead of *Mishap Incident Report*. Use of the terms *"Salvor"* and *"Salvor report"* are not meant to single out the Captain or crew of the *Salvor* as entirely responsible for the Navy's search effort. The *Salvor* did not even arrive in Washington waters until the search was almost over. *Salvor's* commanding officer, however, was the ranking member of the active search, and as such was responsible for the post-search report and its conclusions.

A great many friends and family reviewed the manuscript in various stages of completion and made valuable suggestions. I am sincerely grateful; you know who you are. My deep appreciation to my wife, Laurie, daughter, Emily, and son Jared who once again put up with another of my hare-brained schemes. As I imagine the three of you rolling collective eyes in amused tolerance, I am filled with love and an appreciation for the Hunt sense of humor.

In memory of Ron Akeson, Captain John Ayedelotte, and Eileen Brown.

Dedicated to Parkinson's warriors everywhere.

Peter Hunt
May 2017
Whidbey Island

Helpful definitions/abbreviations:

Synchronicity: According to psychoanalyst Carl Jung, coincidences with no causal relationship that are meaningfully related; "meaningful coincidences."

Dyskinesia: Impairment of voluntary movements resulting in fragmented or jerky motions (as in Parkinson's disease). **Merriam-Webster Dictionary Online**.

Fluid, rhythmic swaying of the torso, writhing of limbs, strained facial expressions, and an inability to speak articulately brought on by an overdose of Levodopa. **The author's personal definition.**

Dystonia: Any of various conditions (as Parkinson's disease and torticollis) characterized by abnormalities of movement and muscle tone. **Merriam-Webster Dictionary Online**.

Painful, continuous muscle contractions of the right shoulder, wrist, and ankle due to insufficient Dopamine; brought on by an undermedication of Levodopa. **The author's personal definition.**

AIS: Automatic Identification System
DBS: Deep Brain Stimulation
FOIA: Freedom of Information Act

JAG:	Judge Advocate General
MDS:	Maritime Documentation Society
NATOPS:	Naval Air Training and Operating Procedures Standardization
ROV:	Remotely Operated Vehicle
TACAN:	Tactical Air Navigation
VTS:	Vessel Traffic Service

List of charts and maps

Page xiv   Washington State waters
Page 4   Shipping lanes and Rosario Strait
Page 45   The search area
Page 118   Radar log flight paths of 510 and 502
Page 200   510's flight path and crash site

Introduction

Parkinson's disease is at best a constant chore. It can also be a life-changing event for the positive. It is the human equivalent of a run-on sentence, following a meandering and listless course across unknown terrain, always searching for continuity and focus. Parkinson's is an existential challenge virtually every waking moment, and not just physically. The psychological impacts of the disease transcend the physical, both good and bad: the search for identity, the attempts to explain without advocating victimhood, the intrinsic confusion of waking at one in the morning unable to walk and then working out like a machine at the gym three hours later.

Learning to eat and drink with head carefully bowed to avoid aspirating food and water. The constant concern of wondering if the rest of the world thinks you are drunk, and then not caring as you shake and sway through the daily routine of work, chores, shopping, the kids' sporting events. Driving home utilizing every skill and ounce of discipline garnered over five decades to satisfy your conscience that you are still maneuvering safely, stone-cold sober but in a rhythm of repeated flow that focuses the mind, but threatens to allow the car to stray if inattentive.

And then appearing before some of the same bewildered onlookers ninety minutes later, apparently "normal," but with a spine-deep fatigue from the day's earlier battles. It's a constant drag that requires tremendous energy and imagination to adapt to utterly different physical states multiple times a day. It's hard. It can also be satisfying and even fun in a distorted way. I have always enjoyed laughter. Parkinson's demands laughter. How can that be all bad?

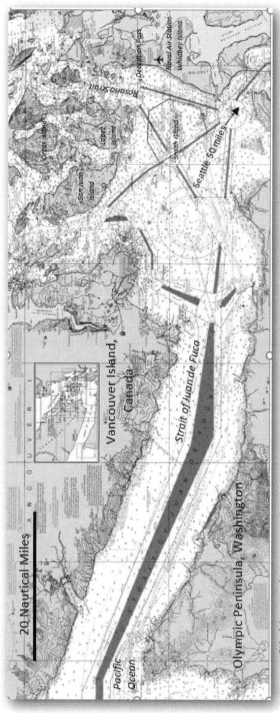

Washington State waters from the Pacific Ocean to the San Juan Islands (NOAA chart adapted by the author).

One

Dawn's patiently eager specter

August 14, 2014

The fog moves with inevitability, not with the weakness of a human trait like confidence, but with the known outcome of the preordained. Slithering into the boat's joints, the fog slowly fills each void until it can no longer be seen. But it is there. Only visible to the mind's eye, it tickles at my ankles with the cold distance of a scythe, sending an icy squirm to the pit of my stomach.

My legs start to tremble, then shake. The legs are still mine, I remind the shapeless cloud, and not politely, as I kick my right foot violently into the emptiness adjacent the helm seat. Ben and Rod turn abruptly at the flash of movement before hastily adjusting their gazes elsewhere.

I can do this. The silent words beat back the fog, if only a little.

I can do this. Still unspoken, but louder in my mind this time. I breathe in deeply, expanding my chest in defiance with a twisted, painful smile.

"I can do this," out loud now. My thoughts; my mouth; my words. I still control where my body goes and what my mind thinks. The mist claws at the back of my head, trying to bend me forward in a bow of submission before entering my ears with a faint ringing that envelopes the brain.

I force my neck straight until my eyes finally reach the instrument displays eighteen inches above the boat's dash. My right-hand edges forward, toward the autopilot controls, and with a determined double push of a button, the autopilot turns the boat starboard two degrees.

Effort is measured in inches and feet and miles, but success is only determined by the will to keep trying. I welcome the fog; it is now a part of me. It is best to keep my enemies close.

The thick outside fog, some would call it the "real" fog, defies depth perception. It is the sort of trick the disease loves to play, giving no hint to where an ambivalent nature surrenders to the persistent malady. Ben and Rod can't see my personal fog inside the cabin, but they must know something has arrived. They can't see the misty tendrils pull at the hooks in my body. All they perceive is my dance of resistance, the effects of the Parkinson's disease: the near-constant writhing and rhythmic sways of dyskinesia, interrupted by dystonia's sudden, spasmodic jolts and tortuous twisting of limbs and spine. I look forward, past the boat's white fiberglass bow, straining eyes for an outside reference point. Nothing.

The curling mist surrounding the boat confuses the senses with nature's subtle perfection, filling each gap to uniformity until the horizon is lost in a rolling sea of gray. Without a skyline, the human eye loses its intuitive perception of physical position; nothing is certain in fog's fake vista. Trying to pierce the veiled obscurity is as if combing a vast, empty asphalt parking lot looking for a penny that is not there. With eyes cast down, straining to see what is impossibly missing, the image becomes increasingly confusing without context.

Accomplished mariners, those who regularly venture out to sea, are wary of fog. They put their trust in navigation instruments, radar and chart plotters, manmade tools that provide substitute realities for the senses. But only the eye's glimpse of a turning ship, the sound of an approaching horn, and the feel of the balancing horizon underfoot can blend seamlessly into a shared awareness. Not even the finest of navigation instruments can compete with the fluid speed of the senses. Instruments do, however, trump the senses in one critical respect: they don't lie. Confused senses don't stop relaying questionable information; instead, they give incorrect information.

The saying of record, whether in the air or on the water, is to "trust your instruments." But there is a corollary to this saying, one I remember well from the steady handedness of a past life as a Naval Aviator: "A peek is worth a thousand instrument scans."

The vibrating rumble of the boat's two diesel engines is loud inside the cabin, but it is even worse on the aft deck where Rod stands between a large winch and a wildly vibrating portable gas engine. Rod visibly strains to hear the horns of approaching vessels. We make brief eye contact, but no message is passed. In the mild claustrophobia of the disease, I fight the urge to read too much into this. Sometimes nothing means nothing.

Ben, the third man aboard, sits at the settee table with open laptop, scrutinizing the imagery transmitted from the sonar's towed array, trailing underwater 700-feet behind the boat. I look left to a computer monitor, its base crudely duct-taped for stability to the fixed galley end table. The track lines of earlier runs still marked on the display define the course I steer. The boat offsets the previous run in a gradually tightening circle, spiraling inward like a snail's shell.

Our search is at the edge of Washington State's San Juan Islands, fifty miles to the northwest of Seattle. There are three main landmasses in the San Juans: Lopez, San Juan, and Orcas Islands. Many smaller isles and semi-submerged rocks dot the charted depths. Despite being well removed from the open Ocean, broad expanses of water in the San Juans allow for surprisingly heavy seas.

Massive oil tankers ply the deep waters, transiting between the open Pacific seventy miles to the west and the nearby Anacortes oil refinery. Freighters and ocean-going tugboats pulling formidable barges also travel these waters, hauling cargo between Vancouver, British Columbia, and Seattle. Our search today is at the confluence of this varied commercial traffic, in a stretch of water known as Rosario Strait.

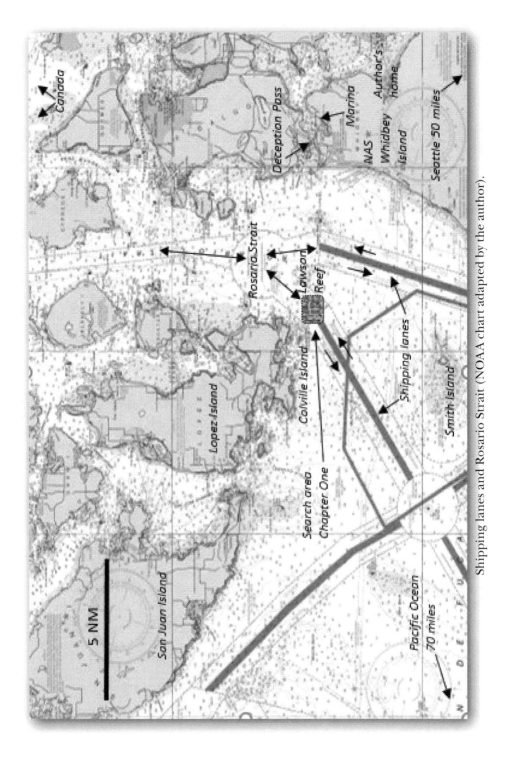

Shipping lanes and Rosario Strait (NOAA chart adapted by the author).

The Lost Intruder

Mixing scores of recreational boats with commercial shipping is always cause for concern. Add bad visibility to the mix, and it can be a recipe for disaster. Summer boaters weave seemingly random paths across Rosario Strait's busy commercial shipping highways, darting between vessels so large that it can take miles for them to turn. Directional traffic lanes separate the opposing shipping in Rosario Strait by a quarter mile, but few rules are uniformly practiced for de-conflicting the myriad pleasure boats that work the fishing grounds. Puget Sound's relatively narrow entrance at the open-ocean complicates matters further, creating a tidal current so swift that, at times, it is impossible to overcome for a slow-moving vessel.

Operating in Rosario Strait in the fog can be a challenge, but most boats merely transit the area and are gone. We are different. Our reason for cruising in the Strait has nothing to do with shipping or fishing or simple pleasure boating. We are searching the ocean bottom for what does not belong. We are looking for a Navy jet.

I've made my share of boating mistakes over the years, many of them on our search vessel, the *"Sea Hunt."* What's troubling is the uncertainty that I've learned a damn thing from such past errors. One lesson believed to be ingrained decades earlier was to only enter fog if given no choice. It isn't enough to responsibly operate a recreational vessel in poor visibility; fog forces a Captain to trust in the knowledge and ability of every surrounding boater as well.

These lessons were learned through experience's harsh litany of hard-to-swallow mistakes and close calls. Pride can lead a mariner to disaster in the blink of an eye, and it is pride that has put us in a precarious position. *Sea Hunt* operates in the narrow shipping lanes with less than fifty feet of visibility, restricted in movement due to the sonar cable strung out behind us, constrained to moving deathly slow. We are vulnerable to fast moving fishing boats and large ships alike, constantly tempting every Captain's worse nightmare—collision at sea.

Sea Hunt's navigation instruments are not acting normally, and the display images match the tremor in my hand with a jitter of their own. The instruments have been acting up with troubling inconsistency all morning, becoming nearly useless the moment *Sea Hunt* entered the fog. So much for trusting my instruments, I think.

Radar is the only sure way to spot other vessels when maneuvering in bad visibility. Much like an above the waterline version of sonar, a radar emits energy pulses that are reflected off hard materials, such as land or a ship. This returned energy is translated by a receiver into a visual image. The time it takes for an energy pulse to make the round trip to a target and back is converted to the distance from the radar. As *Sea Hunt* creeps along its racetrack pattern, Ben searches the ocean bottom with the sonar, while I scan the water's obscured surface with the radar, looking for ships, boats, and buoys. Rod's responsibilities straddle both endeavors, as he operates the sonar tether's winch while looking and listening for danger.

The chart plotter uses global positioning system satellite data to build a map of a boat's progress across the water. To have both radar and chart plotter act so strangely is unprecedented for me; it is confusing as hell. My ability to maintain a mental plot of our position relative to the surrounding vessels is eroding. Situational awareness is slipping from my grasp.

It takes hours before I conclude that the radar and chart plotter are generally accurate. It appears that the swift currents are causing the gyrations of our snail-paced vessel: I've never traveled this slowly for so long in a boat. *Sea Hunt* is restricted to a speed of between 1½ and 3 nautical miles per hour. Any faster and the image being sent up from the "towfish" at the end of its long cable leash will become distorted and unusable. Any slower and the entire cable will droop, and then drop, risking a scrape of the expensive transducer array and its attached communication cable on the sea bottom. Sonar scanning in a strong current makes for a stressful, nearly impossible, workload for me. It zeroes in like a dive bomber on my disease-induced vulnerabilities.

Turning my head, I struggle to raise my voice, "Coming left!"

The words are softly muffled, barely audible. Like trying to play a wind instrument with a severe chest cold, there just isn't enough air in my lungs to make my voice heard. Ben and Rod look forward, their faces blank. I raise a shaking hand and point left until they both nod in acknowledgment. We turn.

The Lost Intruder

I leave most of the critical lookout duties to Rod on the aft deck, a job he takes seriously. He sticks his head in the cabin with a shouted update. Reaching into a small cupboard to the right of the helm, I motion for him to wait, and hand him a portable air horn. Rod is an experienced mariner; an Alaskan charter boat Captain for many years, I trust his judgment. He listens intently from the aft deck, focusing his gaze into the thick blanket of fog, trying to discern the real from the imagined. He shares my concern with the radar blip that was behind us moments ago but is now approaching from the left within a mile and closing fast.

"Pete!" Rod steps through the open sliding glass door leading to the aft deck to speak. "Can I try listening at the bow, get further from the engine noise?"

I weigh the risks. Although the seas are calm, if Rod loses his grip on the side rail and falls, we would be hard pressed to find him. One slip could mean a hypothermic death in the fifty-degree Fahrenheit water in less than two hours. Assuming, that is, that he is not run over by a ship first.

I turn to Rod, "No, let's not chance you falling in. Try the flying bridge instead."

Rod nods in the affirmative, and I appreciate his calm professionalism. Despite having more experience on the water than me, he knows that there can only be one Captain. Argument is not just inappropriate; it can be a deadly distraction. Rod climbs the flybridge ladder, only to return a minute later, shaking his head in the negative. He leaves the cabin door open.

The throttles are in continuous motion trying to keep up with the always changing current, caused by the more than one cubic mile of water that is exchanged between Puget Sound and the Pacific Ocean with each of four daily tides. I take a quick peek out the forward windscreen, straining to see a couple of extra feet. Nothing. I steal another glance aft. Rod points the air horn left of *Sea Hunt's* course, letting out a long blast. Cocking his head to the side, he listens. Nothing.

Turning back to the radar screen, I wait impatiently for the arcing sweep to light up the mystery vessel. The radar blip appears again, this time closer to *Sea Hunt*. The display screen's gyrations finally settle down as *Sea Hunt* steadies up into the current.

"This guy's going to try to pass us," I say to Ben, not expecting a response.

At 38, Ben has proven himself an accomplished explorer as a diver and, now, as a side-scan sonar expert. He is a perfectionist with little tolerance for error. In some ways, he reminds me of myself before my Parkinson's diagnosis, which might explain the tense atmosphere in the boat's cabin: perhaps we each see something familiar in the other that we would rather not. Ben does not say a thing or break from his intent study of the ocean bottom.

Lately, due to my difficulty articulating words, I am frequently ignored. These one-sided conversations are particularly frustrating, cloaking me in a bubble of isolated invisibility. Those who can acclimate to my fidgeting soon reach their threshold of patience trying to understand my softly garbled speech. By afternoon, I am treated by most as a child, and the exchange of invisibility for condescension is the toughest pill of my medication regimen to swallow.

"I'm showing him within a quarter mile and closing," I say, twisting uncomfortably back toward Ben to make eye contact. "Ben, I strongly recommend that you start reeling in the towfish."

Ben glances up, hesitates, and then looks back to the laptop without comment. A few seconds later, he stands to go out onto the aft deck. I have no idea if he understood me. This is the standard communication process for us lately, but I am out of ideas and too damn exhausted to care. I struggle to focus on the radar while fighting through painful contortions that swirl in tight circles on the vertebrae in my lower back. The radar sweep paints the mystery vessel in a blossoming point of light, and again, it is closer.

"This guy is definitely overtaking us," I can barely hear my own voice.

The unknown boat is getting dangerously near. If the vessel had maintained the predictable course and speed of the previous mile and a half, then *Sea Hunt* would have crossed the shipping lanes well ahead of its projected path. But that's now impossible. Why are there so many small boats in the area, I ask myself, each moving erratically? There is no time to dwell on the question. It is only after returning to port that we learn it is opening day for the King Salmon fishing season. This explains the unusual changes in speed, as barely moving fishing boats troll for salmon, and then abruptly sprint for the next fishing spot after the lines are reeled in.

The radar sweep illuminates the mystery vessel again. I bolt upright, eyes glued to the radar screen. The fast-moving radar return is overtaking us in a tight, right turn, moving towards *Sea Hunt's* bow. With the mystery boat now pointing directly perpendicular to our course, our two vessels risk colliding. It makes no sense; does the boat's captain even know that *Sea Hunt* is in his path? We are getting boxed in, unable to turn left with the unknown vessel veering into us. To the right are the shallows of Lawson Reef. I steady up the helm on our original search pattern heading. Rod must share my concern; he lets out another long blast on the air horn.

"He's crossing our bow, less than 200 feet—can you see him?" I try to yell to Rod.

Rod scans the fog intently before shouting back, "Nothing Pete, I can't see a thing!"

The mystery vessel is almost directly in front of us when its radar return merges with ours, getting too close to be depicted on the smallest scale of the display. I sit indecisively for a second, trying to think clearly as my entire body shakes. A loud noise breaks my stasis. The gong of a large bell reverberates from the open window to my right: it is the sound of the Lawson's Reef marker buoy, surprisingly near. That cinches it. I pull the throttles back to idle and put the engines in neutral. The walls close in as the bow wake from the still unseen radar contact begins to reach *Sea Hunt*. I stumble out of the helm seat, reaching for the inflatable life vest on the floor behind me, convinced we are about to collide.

The unknown vessel, whether it was a mid-sized fishing boat or a larger commercial ship, never comes into view. A second later, Ben storms through the open cabin door after seeing slack in the tow line and the shift levers in neutral.

"What are you doing?" He yells, "The towfish is dragging on the bottom. Get us moving!"

Ben rushes back on deck to reel in the sonar array as the wake from the passing vessel pitches *Sea Hunt's* bow in a sweeping arc. I start to speak, promptly think better of it, and push the throttles forward, too much at first. I drop the life vest, slow, and try to get my act together. My eyes scan for a reference, a piece of the horizon, anything to ground my senses: nothing but fog. The afternoon exhaustion overwhelms. Acute anxiety shoots through me, causing my shoulders and legs to spasm violently. I struggle to breathe. It feels like I'm drowning.

Finally, ten minutes later, the last of the tether cable is wrapped around the winch spool that dominates the aft deck. I see with relief that the towfish is still attached. We are clear to maneuver. Ben looks over the cylindrical towfish carefully, occasionally pointing out the damage to Rod. I take a couple of deep breaths, trying to shut out Ben's anger, but my mind's a jumble. We are not out of the woods yet—there are other boats still in the immediate area. I know through my two decades as a Navy and airline pilot that this is the most vulnerable time in any high-stress situation, that it is critical not to respond to distractions. Still, I can't shake it.

Ben is right to be angry. It stings, but it wasn't my $16,000 towfish dragging along the bottom. At the end of the day, even if my interpretation of the unfolding radar events was entirely accurate—which is by no means certain—I allowed circumstance to box us in, leaving no way out. Responsibility lies with the captain. In an epiphany, it is evident that the root cause of the incident is my inability to accept my new limitations. I can no longer handle the complex task loading of low visibility towing, certainly not while crisscrossing the shipping lanes. It was hubris to think that I could. This is my foul up.

Rod is listening and tries to take some of the bite out of Ben's criticism. "Pete, I know a few radar techniques that might be helpful."

I appreciate Rod's attempt to smooth things over, but this is not the time. We are still in the fog, near Lawson Reef, in the shipping lanes. I turn to both men, struggling for composure.

"Okay, listen up—that's enough. We can talk about this back at the dock. Right now, we need to run the boat." It takes all my strength to muster a commanding voice, but it must be done. Embarrassed, exhausted, and nearly beaten, I wonder what the hell I'm doing here. For the first time, it honestly hits home that Parkinson's is going to win.

The Lost Intruder

Ben sits quietly for a few minutes with an intense look on his face: Anger? Hatred? Disbelief? All three? I can't tell. I don't ask. Finally, he speaks.

"We have one small area left to scan. Are you up for it?"

I look at the computer monitor. We have covered the entire day's sector except for a small portion in the middle of the screen. It should take about fifteen minutes to cover the tract.

"What are the chances, Ben, really?" I just want to get the hell back to port.

"Pete, I'll tell you, it might sound like a good idea to leave now, but I've done this before. You'll wake up in the middle of the night and just know that the airplane is in that spot. If we don't cover the area now, we'll have to make a special trip later."

Ben's right. I muster my energy. "Okay, let's finish it up."

The cabin falls into an uncomfortable silence. Rod loiters uncomfortably by the aft door before going back on deck. The towfish shows some scratches, and there is a multitude of cuts along the tether cable that will eventually require repair, but as the winch spool unwinds it appears that the sonar unit is still functioning.

"Hold on, what's this?" There is now excitement in Ben's voice. "Look at this!"

A sizeable contact comes into view on the same monitor that has shown nothing longer than three feet on the flat bottom after days of searching. By comparison, this thing is huge.

"The target is definitely man-made," Ben says confidently. "This might be it!"

"It's in the correct area, that's for sure." I glance over to Ben's laptop. "I don't know; I'll have to take a hard look at it, check some measurements."

After eight months of research and over fifty hours of surveying the ocean bottom, had we finally found the lost Intruder?

Two

Cruel with well-honed diversion

August 2014

I turn left off Washington State Route 20, just beyond a barely visible runway threshold and a series of hand-painted billboards. Each homemade sign sports a stenciled silhouette of a military jet with four words at the bottom written in bold red, "The sound of freedom." The morning calm is abruptly shattered by the high-pitched scream of jet engines as a sleek E/A-18G Growler, just 400 feet overhead, makes a shallow turn toward the runway. There is a slight hesitation before the jet's invisible wake erupts in a deafening roar, forcing hands to ears and physically shaking my Ford Expedition. The whine of jet turbines slowly recedes. In a past life, a much younger me had flown an older, uglier aircraft to the same runway, startling drivers and spooking the cows grazing in the expansive field alongside the road.

The air becomes still and, slowly, the sound of chirping birds returns. I pull into the driveway, get out of the car, and walk through the downstairs entrance to my home office. The door has not yet shut behind me when my eyes are drawn to a sheet of paper lying on the desk. The print out of the underwater sonar image is oddly comforting; it makes me feel grounded. I lean on the oak desktop to stop my body's rhythmic swaying and stare hard at the picture. At first, it doesn't look like a jet. Still, I know that it's the old warbird. My old warbird. Piece after piece of the mystery are falling into place.

Thirty years earlier, I had joined the Navy to fly. There are few occupations as well defined and rigidly understood as that of a pilot; it can be a disconcertingly comfortable identity. After Navy flight school, I had been assigned to fly the A-6 Intruder. The "A" stood for the "Attack" mission, the job of bringing the fight to the enemy as a low-level bomber. Many former A-6 aviators opted to stay on Whidbey Island after they either left the Navy or retired, but none of these men were obsessed with the lost Intruder. What made me different? What possessed me to turn a past association with an old aircraft into an arduous, even a dangerous, quest?

The missing A-6, set apart from hundreds of other Grumman-built Intruders by the permanent numerical designation of 159572 painted on the tail—the bureau number—had been in our squadron for about a year and a half when it crashed. This was more than long enough to develop a black-cloud reputation for mechanical problems. Every Navy jet has two independent identification numbers: the unique six-digit bureau number on the tail, and the more practical, three-digit squadron identification number painted on the nose. The lost Intruder had 510 painted on its nose when it crashed. While the bureau number was a more precise identification, I had come to think of the lost Intruder as simply "five-ten."

After coming off the Grumman manufacturing line, 510 was initially assigned to the Marine Corps in 1978 as a part of the first deployment of attack squadrons to Iwakuni, Japan. In 1981, 510 transferred to the Navy. She then flew from the deck of the Japan-based U.S.S. *Midway* for half a dozen years before returning state-side to Attack Squadron 145 at Naval Air Station (N.A.S.) Whidbey Island. Attack Squadron 145 was my old squadron. Five-ten was one of twelve Attack Squadron 145 Intruders when I checked in for duty, fresh from A-6 flight training, in September of 1988.

According to my Navy log books, I had flown 510 four times: three flights from the aircraft carrier U.S.S. *Ranger*, and one from N.A.S. Whidbey Island. Three of the sorties had been training exercises, and the fourth was as an aerial refueling tanker, one of the Intruder's

A-6 bureau number 159572 in its original Marine Corps paint scheme (circa 1978, photographer unknown).

supplementary missions, from *Ranger* in the Indian Ocean. I had also been witness to the unfolding drama of 510's demise. Standing in the ready room, the squadron's central meeting space, I listened as 510's aircrew first reported a problem over the Squadron Duty Officer's radio. The bombardier/navigator of the Intruder, Lieutenant Chris Eagle, was one of my closest friends. The pilot, Commander Denby Starling, was the squadron's Executive Officer, the second in command. He would be our Commanding Officer a year later during Operation Desert Storm. Denby Starling would go on to have a distinguished Naval career, eventually retiring as a Vice Admiral. Neither man suffered any permanent ill effects from the ejection, a violent, hazardous last-ditch procedure that got them out of the uncontrollable aircraft before automatically deploying their parachutes. The ability to safely eject was the only thing that saved the two from certain death.

What fascinated me the most about the lost Intruder, from the day it crashed on November 6, 1989, was intensely personal. Most of my compelling life experiences have come from flying Navy jets and a life-long hobby of scuba diving. To have a submerged A-6 within miles of my home, one directly linked to personal history, was not just an interesting fact: it demanded action. Just not necessarily urgent action. For 25 years, I would occasionally daydream of randomly running across the missing A-6 while diving, but wasn't so delusional as to think that there was a realistic chance this could happen. Far more qualified experts had conducted an active search for the jet, only to walk away empty-handed.

The Lost Intruder

Attack Squadron 145 nose number 510, bureau number 159572, visiting Andrews Air Force Base in July of 1988 (photo courtesy of David F. Brown).

Soon after 510 plunged into Puget Sound, a search effort involving four Navy ships lasting two full months ended in frustration. The Navy blamed the failure to find the $30 million aircraft on swift tidal currents, deep water, poor underwater visibility, and inclement weather. Tidal currents, caused by the moon's gravitational force, are pushed to extremes in the greater Puget Sound. The tidal ebb and flow of massive volumes of water move through the Strait of Juan de Fuca between the Pacific Ocean and a vast network of relatively narrow, long waterways, creating currents that can exceed eight knots in certain spots. The chance that I would stumble across the jet resided somewhere between extremely slim and zero percent. All the same, it was still conceivable, and it is the possible that often gives our lives purpose.

It was not until late 2013 that I finally decided to look for the missing jet. The lost Intruder began to represent a neat and tidy wrap-up of my past that I could no longer deny. It had taken nearly 25 years to develop the nerve to entertain the notion that I might succeed where the U.S. Navy had failed. But it wasn't just the clock ticking behind me that got the project moving forward; it was also the countdown of those hands into the future. Parkinson's disease was eroding my capabilities with each sweep of the second hand. The search for the lost Intruder had turned into a now or never proposition. I chose now.

I was working as an airline pilot, having just turned 43 years old, when diagnosed with early-onset Parkinson's disease in 2005. Being a pilot did not make the transition from "normal" to "something other than normal" any easier for me. After ten years of walking into the United Airlines Flight Operations Office in Seattle feeling steady as a rock, early one morning my hand started to shake uncontrollably. There was no warning, no identifiable foreshadowing clues. It was just me—coincidentally speaking with my boss, the Chief Pilot for Seattle—when my right hand decided to suddenly disobey a lifetime of neurological commands. It was like turning on a switch, and in an instant, my life changed forever.

At first, I didn't know what to do about the tremor. I hid my hand in my jacket pocket and tried to figure it out on my own for a month. I got nowhere. I called a friend who is a medical doctor. I called the pilot's union. They both advised, "Get thee to a neurologist." (The union attorney said those exact words.) And just like that, my twenty-year flying career was over. The change was abrupt, but I adjusted as the Navy had trained me to adapt to new situations. I accepted the facts as they were, even though in retrospect, I had little understanding of what those facts ultimately meant. I prioritized. I did my best to put on a good face to insulate my wife and two children from the coming sea change.

By Christmas 2013, I needed to act if the Intruder search was to move from the "someday" category of dreams to the "at least I gave it a shot" column. The debilitating effects of the progressive, neurological disease were becoming too severe to wait any longer. Deep Brain Stimulation (DBS) surgery was scheduled for the fall of 2014, a final effort to slow, and maybe reverse, the disease's progress. This gave me ten months to find the missing jet before my underwater adventures would be severely curtailed. A successful DBS procedure meant a scuba diving limit of 33 feet. If the DBS surgery didn't work, then my condition would not allow further exploration anyway. In any case, having run out of 25 years of reasons not to look, I did all that was left—I started to search.

In many ways, looking for the lost Intruder offered a substitute for a closure that I fully realized would never exist in real life. Discovered truths tend to branch out in unexpected directions, creating multiple new questions. My battle with Parkinson's did more than instill in me a hope of finding the jet, it fostered a profound belief that anything was possible if I honestly gave it my best effort. And if the lost Intruder couldn't be found in the end, however that "end" might be defined, then I would know for certain that it was unfindable, at least by me.

I had dived many sunken ships in my life, but had never searched for an undiscovered wreck and had no idea how to start. Diving has been an important part of my life for over 35 years. It has affected my personality in ways both subtle and considered and is nearly always running in the background clutter of my thoughts. Diving inspired the name of my boat, a play on words with my last name and the popular 1950s TV show "*Sea Hunt*," where Lloyd Bridges thrilled viewers with weekly underwater adventures. Before joining the Navy at age 23, I had been an accomplished wreck diver, making 13 dives to the Mount Everest of shipwrecks, the *Andrea Doria*. These were not gentle sight-seeing dives. These were among the first deep penetrations of the *Andrea Doria's* interior by anyone on scuba, hazardous dives that would become synonymous with deadly accidents in the decades to come.

Although I didn't know how to start the physical search for the lost Intruder, I had been to college and knew how to conduct general research. It made sense to narrow down the potential crash area as much as possible before striking out on the water to look for the missing jet. In January of 2014, in Washington's dreary winter gloom, the search for the lost Intruder began.

There were practical reasons for starting the project with extensive research. Although I owned a boat, I had no sonar, metal detector or any practical method of surveying the ocean bottom. With an incurable illness, no prospect of financial reward, little chance of success, brain surgery looming, and one child in college with another about to start, I was not in a position to spend thousands of dollars on a search. Still, desperate for a distraction, anything to pry my focus away from the disease, I decided—the hell with Parkinson's. I'm doing it. Trusting that time would sort out the details, I began to research 510's crash while experiencing an unearned confidence that events would somehow fall into place. And with that shaky premise stiffening my twisted spine, I was on my way.

My family had adjusted to much in recent years, and I assumed that they would take my new project in stride as if another symptom of the disease. Or, maybe, they would consider it simply one more of Dad's life-long habit of eccentricities. Either way, I never asked my wife, Laurie, or my children, Emily and Jared, what they thought about my quest until it was all over.

Fortunately, there were two documents long filed away that might shift the chances of finding the crash location to within a needle-in-a-haystack range of probability. Both papers were from the official Navy record and were given to me by my friend from decades earlier, Chris Eagle. As the bombardier/navigator of 510, it was no surprise that he had also entertained someday mounting an effort to find the jet. But the Navy moved Chris to the post-graduate school in Monterey, California in the early 1990s, and his casual daydream had little realistic chance of moving forward. Chris Eagle would never be stationed at N.A.S. Whidbey Island again.

The first document was Chris Eagle's handwritten statement outlining the events leading up to the accident. It was dated November 7, 1989, the day after the ejection. The single-spaced, three-page statement was not an easy read, even for someone well versed in Navy acronyms. After spending decades listening to Chris talk about the incident, it was also difficult to separate his slowly evolving memory from the perceived facts of the statement. The charismatic power of an oral history increases over the years as a tale is told and retold, with each nuanced change reinforced by positive audience feedback until it becomes folklore. The written record tends to be dull, stuck as a static snapshot of apparent reality.

The second document was typed on official Navy letterhead. Document two was the fourteen-page post-search operations report of the Navy's dive and salvage ship, the U.S.S. *Salvor*. A minesweeper from nearby Naval Station Bremerton had been the first Navy vessel to reach the vicinity of the crash site on December 27, 1989, but, in January of 1990, the Captain of the *Salvor* was assigned as leader of the recovery effort. Despite meticulously searching the area with sonar, the minesweeper had no luck in finding the stricken jet. From mid-January to mid-February, four separate U.S. Navy vessels searched the area in an expanding grid that eventually encompassed 29.3 square miles of the sea bottom. Utilizing side-scan sonar systems, minehunting sonars, and several different types of tethered, unmanned mini-submarines, the Navy effort found and investigated several promising contacts. None of them turned out to be the missing jet.

U.S.S. *Salvor* (photographer unknown).

Enclosed in the report were a chronology of events, the names of the participating Naval units, and a brief summary of the search's progress and the difficulties encountered. Rounding out the report was a rather lengthy list of the high-tech underwater search and diving equipment utilized, of which I had none. The chronologic narrative of each day's activities told a familiar story for the region: swift currents and stormy winter weather delayed the search again and again. When the *Salvor* weighed anchor to head back to Honolulu, the Navy ships had conducted a combined total of eighteen days of sonar scanning and six days of Remotely Operated Vehicles, or "ROV" operations. ROVs are small, unmanned submarines equipped with cameras for underwater identification purposes. The material cost alone in consumables, such as fuel, was $34,758.73. Inflation-adjusted for 2014, the price tag came to $66,784.41. This didn't cover a penny of the cost to build or maintain the ships or equipment, nor did it account for the pay of the hundreds of sailors involved in the search.

The *Salvor* report also listed the latitude and longitude coordinates of the twelve search grids deemed to have any probability of containing the lost Intruder. Six square miles in this area were annotated as the high probability search sector. The location of the grids was based on a centralized ejection position. The Navy had taken an estimated ejection point, drawn concentric circles out to several miles, and then covered the entire area in a rectangular grid system. The report stated that the ejection position was based on a distance and heading from a navigational aid at N.A.S. Whidbey Island, but the report gave no indication as to the source of this information. The lack of sources in the report bothered me. What if the *Salvor* had gotten the ejection position wrong? Might that be why the Navy's search was unsuccessful?

I sat back at my desk to consider the challenge. I had first-hand knowledge of Naval Aviation standard operating procedure, as well as the flight characteristics of the Intruder. As a former A-6 instructor, I was also aware of how Navy pilots and bombardier/navigators were trained, and what might be going through their minds during an emergency. The nexus of unknowns congregated at the location of the ejection. I decided to approach the problem with the assumption that the *Salvor's* assumed ejection position was incorrect. The downside to this method was that it left me with precious few undisputed facts.

Once the ejection point was established, the key to narrowing the search area would be developing a sound model of 510's final phase of flight. Chris Eagle's statement implied that the ejection occurred at 6,000 feet, a full nautical mile high. An unmanned jet could fly a long way in the time it took to descend from 6,000 feet. What was the aircraft's airspeed and heading at ejection? Was the landing gear down? Were the flaps and slats extended? The aircraft became uncontrollable due to a complete loss of hydraulic pressure; the A-6 flight controls could not operate without hydraulic fluid coursing through its airframe. How might this have affected the final seconds of 510's time airborne?

The Lost Intruder

Five-ten's energy state at water impact would tell a lot about the Intruder's physical condition resting on the ocean bottom. Did the jet strike the water at a relatively steep dive angle and high airspeed? If so, that might have caused the aircraft to break apart, possibly in tiny pieces, which did not sound implausible given the fact that the Navy's search turned up exactly nothing.

On the other hand, a shallow, slow speed water entry would probably have left the Intruder intact. It might have drifted with the current for a while, or perhaps even "flew" in the water column for an indeterminate period of time. How strong and from what direction was the current flowing when the mishap occurred? Did 510 float for a time, allowing the prevailing winds to push the Intruder before going under? The long list of scribbled questions was troubling. How could I possibly succeed 25 years after the U.S. Navy had failed, back when the evidence was fresh?

I took a deep breath. The Navy had concluded that further search for the missing A-6 was "futile and cost prohibitive." I disagreed. My confidence was based on three factors. First, Chris Eagle had maintained for years that the Navy's search had been conducted in the wrong area. With the Navy's hunt covering such a large expanse, there couldn't be much of the sea floor near the ejection point left unexamined. If I could identify this area, then I might be able to do an end run around some of the unknowns. The second cause for hope also came from Chris Eagle's statement, which answered some of the flight profile questions. Chris's statement did not, however, identify 510's exact position when the crew ejected.

There was one other piece of encouraging evidence in Chris's description of events. Almost immediately after pulling the ejection handle and breaking through the Intruder's canopy glass, as the A-6 ejection seat was designed to do, Chris saw the mortally wounded Intruder flying below him. The parachute's opening shock then yanked him back so forcibly that it caused Chris to gray out. When the stars cleared from Chris's vision, the first thing he saw was the churning bubbles of the sinking jet between his swinging feet. He had been an eyewitness to the most critical piece of information needed to find 510—the lost Intruder's point of water impact.

He came tantalizingly close to identifying the exact location of the sinking jet. Chris Eagle's statement indicated that the impact point was "¼ to ½ mile south of Lopez Island." There was a problem, though. The Lopez coastline was 3½ miles wide at the Island's southern end. Which part of Lopez Island had Chris used as a reference? I decided to test Chris's memory by asking him directly. He replied to my email the next day:

Thursday, February 27, 2014

Pete,

I still can't find my folder (with the ejection files) here in the house. If I had to guess through 25 years of haze, I would say the A-6 hit somewhere between ¼ and ½ mile south of Colville Island, but that is a huge guess at this point.

Chris

Colville Island is a small Washington State Wildlife Preserve about a half a mile south of the southeastern tip of Lopez Island. If Chris's distance estimation and memory proved accurate, then determining the exact ejection point wasn't so critical. This narrowed down 510's likely crash site to about one square mile. It seemed too easy. Why would the Captain of the *Salvor* ignore Chris Eagle's statement?

There might have been a legitimate answer to this question. The *Salvor* report made a note of a test flight in the A-6 simulator to evaluate 510's post-ejection flight characteristics. The A-6 simulator had been set up to mimic the exact flight and configuration parameters of 510 at the time of the ejection.

The test determined that 510 could not have flown for more than one mile before crashing. For Chris Eagle's distance estimate from Colville Island to be correct, the A-6 would have needed to fly at least two miles from the *Salvor's* posited ejection position. If 510 had somehow hit the water further from the *Salvor's* ejection point than the simulator test deemed possible, then the actual crash zone was unlikely to have been thoroughly searched. Why bother if the aviation experts claimed that the jet could not possibly have flown that far? But this all hinged on the *Salvor's* ejection position being correct.

The Lost Intruder

I pulled up a nautical chart of the area on the internet and studied the map's depth contour lines. About 80% of the Lopez Island coastline was within the outer limits of my Parkinson's adjusted personal depth range for diving. If I trained and pushed myself a little, I could probably dive most potential sites close to Lopez.

Soon after my diagnosis, I had researched the disease's limitations on scuba diving. Advice was hard to find, and I took the dearth of guidance as a cue to figure it out for myself. One of the reasons for the scarcity of advice for those afflicted with Parkinson's is that personal limits can vary widely. As the disease advances, existing symptoms become more pronounced and new ones are added, but a person can still look relatively healthy much of the time, despite feeling awful. Parkinson's can also manifest itself quite differently in each patient. The end result is that for many activities, like diving, the individual is on their own in determining what is safe with their particular manifestation of the disease.

Diving with Parkinson's would mean exploring untested limits. There were no courses to teach deep wreck diving when I first became certified in 1978, and it was refreshing to once again anticipate exploration unsullied by training sessions, rules, and etiquette. The only Parkinson's-specific advice I could find was to stop all technical diving immediately. Technical diving is loosely defined as any dive exceeding 130-feet deep or requiring the controlled ascent of decompression. I smiled at the prospect of ignoring the untested recommendation.

Ultimately, it was the inner reaches of the mind, a primal exploration of the unknown, that appealed to me. In the past, shipwrecks had been the vehicle for this self-exploration. Now, it was the lost Intruder. The difference came from Parkinson's, which made self examination now both more urgent and complex. When exploring shipwrecks, I always had a general notion of what I might discover. I had no idea where the search for the missing A-6 would lead.

The prospect of finding, and perhaps diving, the lost Intruder was exciting. It reminded me of my third cause for confidence: I have always been an optimist—I know no other way.

Three

MASKING A STRANGER'S TRUE NATURE

February 2014

Once the Navy gave up the search, there was no compelling reason to look for the missing A-6 in the years that immediately followed. Intruders were hardly novelties, not with at least one operational A-6 squadron assigned to every U.S. aircraft carrier at sea. But by the mid-1990s, the U.S. Navy began retiring the A-6 fleet, and the aging warbirds started showing up at museums. My last time at the controls of an Intruder was in the spring of 1995 during a fifteen-minute flight from Whidbey Island to Boeing Field, the home of Seattle's Museum of Flight. Twenty-five years later, the A-6 was still on display looking no worse for wear.

When Denby Starling and Chris Eagle ejected from 510, the world was vastly different from today. In January of 1989, George H.W. Bush was inaugurated President. The Dow Jones Industrial Average had just reached 2,256, recouping all 508 points lost from the 1987 stock market crash. It was the year Johnny Bench was elected to the Baseball Hall of Fame, and Pete Rose was meted a lifetime suspension from baseball for gambling.

A nuclear Armageddon fought by thawing Cold War foes still held a primacy of fear over that of a warming planet. The Soviet occupation of Afghanistan had just ended. That same year, Chinese troops massacred nearly one thousand student protesters in Tiananmen Square. It was a busy time for the U.S. Navy. Carrier-borne F-14 Tomcats shot down two Libyan jet fighters over the Mediterranean in 1989. There was also a gun turret explosion on the battleship U.S.S. *Iowa,* killing 47 sailors. Three days after 510 plummeted into Rosario Strait, crowds of Germans began dismantling the Berlin Wall. Six weeks after Denby Starling and Chris Eagle ejected, U.S. troops invaded Panama, ousting Manuel Noriega from power.

In November of 1989, I was mid-way through a three-year tour assigned to Attack Squadron 145. Having recently completed my first deployment, this one to the Indian Ocean onboard the U.S.S. *Ranger,* I would eventually ship out two more times before leaving the military for a career with the airlines. My second cruise was to the Persian Gulf on *Ranger* during Operation Desert Storm, and my final deployment, this one on the U.S.S. *Kitty Hawk,* was spent loitering off the Korean peninsula. When Intruders filled the Whidbey Island skies, it was easy to delay the search until some ill-defined point in the future, secure in the knowledge that the A-6's rich history guaranteed interest in the aircraft for long after her star as an operational aircraft went dim. There was no rush.

510 landing on the U.S.S. *Ranger* (from the author's collection).

The bulbous-nosed Intruder was designed to penetrate the Soviet Union's Cold War erected integrated air defenses. Integrated air defenses are a series of overlapping defensive measures, typically fighter jets, surface-to-air missiles, and anti-aircraft artillery. Integrated air defenses guarded the perimeter of the Iron Curtain. They were also the first line of defense for client states flush with Soviet weaponry in other regions, notably Syria and Iraq. Traditional high altitude bombers were expected to have a dismal survival rate when flying through these defenses, and low altitude tactics were devised to compensate. Intruder crews trained to fly to the target at high speed, within several hundred feet of the ground, day or night, in any kind of weather.

The all-weather, low-level attack mission was not for the faint of heart. Although reduced, the danger of being shot down by a ground-based missile was by no means eliminated. Evasive procedures needed to be employed to break a missile's radar-guidance lock, including a combination of high "G"—one "G" representing the force of gravity—turns in conjunction with decoy tools, such as metal confetti chaff and high-temperature flares. Flying low to the target also introduced a vulnerability to small arms fire. The aggressive flight profiles required to avoid these threats exposed the enemy's most basic defense: distracting the A-6's crew to the point where they accidentally flew into the ground.

During the six-week air war of Operation Desert Storm, four of the five Intruders lost were casualties of the low-level environment. Scores of allied low-level attack aircraft, notably British Tornado bombers, were also brought down by either small arms fire or inadvertent flight into terrain. The high mortality rate of the all-weather, low-level mission eventually persuaded the Navy to forgo the tactic altogether. This, in part, led to the A-6's eventual retirement.

West Coast-based Intruder crews trained on numerous low-level routes crisscrossing the Cascade Mountains of Washington and Oregon. A typical training flight would position a lone Intruder fifty miles out to sea to mimic a carrier based night-attack. After crossing the shoreline at 500 mph, the A-6 crew would weave a path down valleys and around mountains for 45 minutes before reaching the target range near Pendleton, Oregon. The first run at the bulls-eye would be painstakingly planned and executed to release weapons plus or minus ten seconds from an assigned target-time.

The A-6 straddled the tactical and strategic roles, compelling crews to train for a vast array of missions. These included the delivery of nuclear weapons, close air support for Marines on the ground, mining waterways, and the use of precision, laser-guided bombs to destroy high-value targets. The A-6 was a first generation surgical-strike bomber, a term initially used during the Cuban Missile Crisis that implied a far cleaner depiction of combat than reality allowed. It is hard to imagine anything "surgical" about dropping tons of high explosives on people. The Intruder was not a sexy jet—crews joked that it was the only airplane built with the "pointy end" in the back—but the A-6 consistently got the job done. It also bridged the gap between short-range fighter-bombers and the Air Force's giant B-52. Second only to the B-52 in payload capacity, the Intruder could carry 28 general purpose, 500-pound bombs compared with the B-52's load of 50.

The innovations that made low-level navigation and bombing possible in the A-6 were a ground mapping radar interfaced with a 1960s vintage inertial navigation system. The "inertial" was comprised of a set of gyros that registered every force put on the jet. A computer would subsequently translate these inputs into position information which, once compared to the radar and a chart, enabled the crew to navigate around the surrounding higher terrain, effectively masking the Intruder's progress to enemy radar.

Pilots and bombardier/navigators flew training sorties together as a regular pair to help develop crew coordination during the attack mission's high task loading. The unusual side-by-side seating arrangement allowed the pilot and bombardier/navigator to see hand gestures and body language during flight. This enhanced communication immeasurably and helped make the A-6 pilot and bombardier/navigator the most effective aviation crew in the U.S. Navy.

Real world testing was accomplished by Intruders outfitted with temporarily installed lasers in the final years of the Vietnam War. In 1979, a forward-looking infrared targeting sensor was added to the underside of the Intruder's fiberglass nosecone. This metal ball housed a thermal sensor and a targeting and ranging laser, enabling the A-6 to drop precision laser-guided bombs. But the Intruder was beginning to show her age in a dangerous way. The jet's wings were developing cracks due to years of G-force loading. Unfortunately, the cracks were only discovered after a tragic in-flight incident that led to the deaths of several aviators after a wing completely separated from the Intruder's airframe.

The wing cracks came as a surprise to the Navy. With no immediate replacement for the Intruder, stop-gap measures were taken to extend the wing life to buy time for a permanent fix. In the late 1980s, Intruders started being flown to a private contractor in Saint Augustine, Florida for re-winging. In conjunction with the addition of the new, composite wings, a final armament retrofit was also incorporated into the Intruder's airframe through the Systems Weapons Improvement Program or SWIP. These most advanced A-6s could fire the newest weapons in the military's arsenal. SWIP upgraded A-6s had at their disposal the High-Speed Anti-Radiation Missile (HARM) designed to destroy enemy missile radars, the Harpoon anti-ship missile, the Standoff Land Attack Missile (SLAM), and the tank-busting Maverick laser guided missile.

The retrofitted Intruders were also upgraded with a vastly improved electronic warning system to alert the crew if their jet was being tracked by enemy radar. Having only recently come from service in Japan, it is highly unlikely that 510 underwent the maximum longevity investments of the SWIP retrofit. Despite having flown 510, I couldn't remember for certain whether the jet had been upgraded or not.

The A-6 had been the Navy's workhorse during the Vietnam era, and 84 Intruders were lost during the war. The Intruder also flew combat over Libya and Syria in the early to mid-1980s, and a few years later against Iranian ships while protecting re-flagged Kuwaiti tankers. The A-6 operated from six different aircraft carriers during Operation Desert Storm in 1991. Intruders were ordered into final combat by President Clinton in the mid-1990s, this time in retaliation for Iraq's refusal to allow International Weapons Inspectors unfettered access to sites suspected of harboring weapons of mass destruction.

But time was winning a long battle with the Intruder. A-6 maintenance was becoming more challenging and costly with each additional year of service, and the old, subsonic airframe—virtually unchanged since the opening of the assembly line—was becoming increasingly vulnerable to modern air defenses. Re-winging continued until the last Intruder squadron was retired on February 28, 1997. When the A-6 fleet was officially decommissioned, about a dozen Intruders awaiting new wings in Saint Augustine were emptied of fluids and sunk offshore to serve a final mission as an artificial reef for divers and fishermen.

When Commander Denby Starling and Lieutenant Chris Eagle were forced to eject from 510, the thought of retiring the A-6 was not yet considered a serious possibility by the politically isolated West Coast aviators. The expansive country acted as a geographical buffer between the Pentagon in Washington D.C. and the operational Intruder crews on Whidbey Island. Avoiding orders to the Pentagon was a mark of pride among A-6 aviators who aspired to stay at the "pointy end of the spear" of America's defense. The Intruder community's lack of political clout in D.C. was an additional factor in the jet's eventual retirement.

A pilot's time in a fleet squadron consisted of an intense eight-month training cycle, followed by a six-month deployment on an aircraft carrier to a distant world hot spot. Rest and regroup for several months, and the cycle would be repeated. After three years of sea duty came three years of shore duty, which for me was an abbreviated tour as an instructor pilot at the West Coast A-6 training squadron. I finished ten years of active duty with a final sea tour. I completed a full training cycle in my new squadron before deploying off Korea during the nuclear weapons crisis of 1994. Six months after returning stateside, I was a civilian filling out airline applications while getting used to the company of an eleven-month-old daughter. I had shipped out on my final deployment when Emily was only one week old. United Airlines was hiring pilots, my interview and simulator testing went well, and I reported to the training department at the recently decommissioned Stapleton Airport in January 1996.

One aspect of Navy flying that would never be replicated in the civilian world was carrier operations. I had manned the controls of 510 during four catapult shots and an equal number of carrier arresting landings, both dynamic events of such intensity that simply thinking about them made the pulse race. Carrier landings, particularly at night, were the most challenging and dangerous routine flying conducted by the aviators of any military service. To have flown the lost Intruder in this environment created a special bond between man and machine.

Most conventional aircraft carriers in 1989 had four steam catapults on their deck. A catapult ran the length of the bow on both the left and right sides of the carrier, forward of the superstructure. An additional pair of catapults straddled the angled deck landing area. A catapult launch was essential to accelerate a carrier jet to flying speed. The catapult shot itself was not a particularly demanding procedure for a pilot, in fact, the crew, in some respects, was just along for the ride. The transition from the catapult shot to flight, however, was another matter. Flying skill was tested at the end of each catapult stroke, particularly if accompanied by any sort of takeoff emergency.

Just getting an A-6 from where it was parked on the flight deck to the catapult for launch was a complex and potentially dangerous task. The pilot would follow a yellow-shirted taxi director's finely choreographed hand signals, often bringing the aircraft within inches of other jets or the deck edge, until the Intruder was aligned with one of the four catapult tracks. These tracks ran nearly half the length of the ship. The reinforced nose gear would be attached to the catapult's shuttle with a drop-down linkage that fit into the catapult track. Once verified secure, the pilot would receive a hand signal to go to full power, trusting that a holdback fitting behind the nose landing gear would keep the jet in place. The holdback fitting was engineered to resist the force of two jet engines at full power but break with the additional energy of the catapult stroke.

After the Intruder was connected to the catapult, a half dozen mechanics would dart out of sight below the edge of the canopy. Scrutinizing the A-6 for proper flight control operation, these launch troubleshooters would give the jet a final look over. The pilot would then signal readiness to launch with a salute. Cupping the flight control stick loosely with his free right hand, the pilot would keep his left hand firmly on the throttles and the stationary catapult grip. This ensured that the throttles were not inadvertently pulled back by the G-forces of launch. Once sufficient airflow crossed the control surfaces, the stick would "come alive," moving back into the pilot's waiting grasp. The Intruder would accelerate from 0 to 150 knots—180 miles per hour—in 2 seconds. In the time-compressed blink of a tunnel-visioned eye, the A-6 went from a standstill to flying a scant fifty feet above the waves.

510 on *Ranger's* flight deck. Note the upright jet blast deflector behind Catapult 1 (from the author's collection).

Landings on the aircraft carrier were even more intense. The arresting gear consisted of four thick steel cables strewn across the angled flight deck about forty feet apart. With the aid of a visual lighting system, called the "meatball," to tell a pilot whether he was high, low, or perfectly tracking a predetermined glide path, the pilot would make constant, small corrections. The goal was to cross the back of the ship perfectly lined up at a precise airspeed and in a steady descent. If done with finesse and skill, the violent touchdown of a carrier landing often came as a surprise. Carrier pilots never "flared" their aircraft in the soft touch down of an Air Force or airliner landing, but instead drove the jet at a steady rate of descent onto the ship's deck to avoid missing the wires.

But there were many pitfalls. If the Intruder were flying a little too fast, the tail hook at the back of the jet would lift slightly, possibly missing all four wires. This was why carrier pilots aggressively came to full power the moment that the wheels hit the deck. If the hook caught a cable, the jet would be stopped regardless of the power setting. But if all four wires were missed, called a "bolter," then the Intruder's engines would already be spooled up at full power and ready to accelerate for flight. Conversely, flying just a few knots too slowly would put the jet perilously close to stalling, a condition where the wing lacks the lift to maintain flight. Once a wire was hooked, and the aircraft came to a standstill on the deck, the pilot would pull the throttles back to idle, raise the tailhook, and taxi the Intruder out of the angled landing area.

The most challenging aspect of landing on the ship was the smooth integration of three landing parameters: glide path, centerline, and proper airspeed as depicted precisely by an angle of attack lighting display in the cockpit. The carrier landing mantra, repeated endlessly until rote, was "meatball, line up, angle of attack." Changing one parameter required a slight adjustment of the other two, a never-ending process of fine tuning until the wheels hit the steel deck. Mother Nature's contributions of inclement weather and heavy seas increased the landing challenge by moving the ship in all directions, requiring even faster, more precise corrections. Regardless if one were tired, hungry, or just having a bad day, there was only one runway available and the environment was totally unforgiving.

The Intruder's rugged construction made me wonder how 510 had held up underwater for a quarter century. It struck me that the passing years might have treated me better than the submerged A-6. Immediately after my Parkinson's diagnosis, the FAA rescinded my Air Transport Pilot certificate, barring me from flying commercial aircraft forever. Subsequently, I lost my job with United and went on company-funded, long-term disability. When first diagnosed, my symptoms were barely noticeable: a tremor in my right hand, rigidity, and a propensity to get cold quickly. Parkinson's affects each person differently, but the menu of possible impacts as the condition worsens are understood reasonably well.

Parkinson's is commonly defined by the inability of the brain to produce Dopamine, a substance that facilitates the electrical signals for movement between the brain and the body's muscles, but also plays a significant role in mood and emotion. The varying Dopamine levels induced by the disease can often be recognized with surprising accuracy. For me, a slight overage in medication significantly increases my propensity for the writhing, uncontrollable body movements of dyskinesia, but it can also temporarily ease Parkinson's deadliest effects: deep depression and apathy.

Parkinson's weaves its way into the fabric of one's life, leading the afflicted down a narrow corridor of choices into an increasingly isolated existence. Feeling perpetually fatigued, it can be a frightening prospect to keep up one's end of a conversation, never mind obligations, as a friend. By 2014, I both dreaded and yearned for the most routine of socializing, a situation made worse by rarely having the energy to even answer the phone.

The most effective drug in countering Parkinson's symptoms—there are no pills to mitigate or cure the disease itself—is Levodopa, which is essentially artificial Dopamine. But it does not work forever. By 2014, Levodopa's effectiveness in me had been reduced to where it was nearly impossible to find the correct dosage balance, of feeling "on," for any length of time. This was despite splitting pills in half and taking metered doses, around the clock, fifteen times a day. I began simultaneously experiencing symptoms of both under and over medication. These conflicting symptoms incited the rapid onset of violent physical transitions that were unnerving to watch, never mind experience.

In the space of forty minutes, I could go from displaying wildly fluid bouts of dyskinesia, to ten minutes of a near normal appearance, only to have my voice trail off as my right leg and wrist locked up in the painful twisting clench of dystonia. The dyskinesia would moderate, replaced by muscular rigidity and joint pain, but never entirely disappear. After another ten-minute pause, my medications would catch up, and the process would reverse. The cycle might repeat a dozen times a day; every day.

Facing the hard truth, I knew that Parkinson's would eventually win—it always did. After diagnosis, I was fortunate enough to begin accepting this fact, which left me to make the most out of my life without pinning my emotions to false hopes for the future that were beyond anyone's control. My daily goal was simple: to deny Parkinson's its ultimate victory for just one more turn of the sun.

That did not, however, make things easy. Gradually losing the ability to partake in physical activities enjoyed for years, yet bored and yearning for adventure, it was the same nightmarish frustration experienced as when running in sand. I had never felt so trapped, both by mind and body. It was time for drastic action. I began the eligibility screening process for Deep Brain Stimulation surgery. Not all Parkinson's patients are suitable candidates for DBS, and comprehensive testing is required before a surgeon will risk performing the procedure.

Following a battery of physical, cognitive, and psychological tests, I was approved for the procedure. I had ten months to find and hopefully touch the lost Intruder before DBS would restrict my diving to shallow waters forever. The surgically implanted DBS system limited scuba diving to two atmospheres of pressure, just 33-feet deep, which encompassed but a tiny sliver of Washington's coast.

The challenge of looking for the lost Intruder provided the opportunity to take on a long-contemplated project that motivated me when the listless apathy seemed insurmountable. Much as the future prospect of flying from the deck of an aircraft carrier or exploring deep inside a shipwreck had awed my adolescent mind, the possibility of finding the lost Intruder teetered brightly at the edge of my reality. I am a stubborn son of a bitch; I know that well. Jumping in with both feet would trap me in a posture of action, forcing me to move forward. It would help me avoid a Parkinson's induced stasis that might just mean a slow death.

It was a ju-jitsu of Parkinson's most debilitating attack; turning the trap of depression into a set psychological path of positive activity. It would also inspire me to explore the inner reaches of my mind in a discovery of self, a process that I was ordinarily too damned exhausted to even consider. And it tied my life together in a way nothing else could. I had flown 510. I had made deep wreck dives before. These need not be the invisible ghosts of past memory; these could be conjured up again in the totality of their original excitement and adventure.

I could be an explorer again.

And, with luck, the project would see me through to brain surgery in October. I loved the water, and the thought of spending most of the summer on the ocean with a clear purpose made the future exciting, a sensation not felt in a long, long time. When it came down to it, there was no other choice: I needed a buffer between me and the edge of despair.

At the time, I had no idea how close to the emotional and physical edge I would come. It turned out to be one hell of a ride.

Four

A FLUTTERING STRING OF CONFUSION

March 2014

The concrete facts surrounding 510's crash were sketchy at best. Lacking complete confidence in my research sources, I decided to email Denby Starling and ask him directly about his recollections of 510's water impact point. To my surprise, the recently retired Admiral responded within hours.

Tuesday, March 25, 2014

Pete,

It is great to hear from you. I will send you a Google Maps screenshot of where the jet roughly hit. We were in a left-hand orbit in the vicinity of Smith Island, but the thing that prompted our ejection was as the second set of (hydraulic) pumps failed, and I tried to turn left, the airplane rolled right, not aggressively, but enough to let me know that it was not responding to stick input. I told Chris, "We're going to have to get out" when that happened, and next thing I knew, he was gone.

I saw the jet briefly before it hit the water. We were at landing speed for the last part of the flight (flaps down electrically, trying to get the gear down) and I think that once we jumped out, the plane pretty much nosed over and went down at a reasonably steep angle. I thought it hit just a few miles south of Lopez Island.

The Lost Intruder

Marjy and I will keep you and Laurie in our thoughts going forward.
All the best,
Denby

"A few miles south of Lopez Island" was a lot of open water removed from Chris Eagle's "¼ to ½ mile south of Lopez" estimate. I frowned, raising a mug of black coffee to my lips. There was an hour until the local public library opened, and I decided to spend the time re-reading the Navy's account of the search. The *Salvor* report cited no sources, leaving few leads to pursue in greater detail. Still, I held out hope that there might be a tie-breaking clue to help choose between the two competing aircrew narratives. If not, then my only other idea was to scour the *Whidbey News-Times* microfiche archives in the hope of determining which account, if either, was correct.

I compared Chris Eagle's rendition of the facts with that of the *Salvor* report. Attached to Chris's email was a Google Earth computer screenshot picture of a rectangular box, marking the area ¼ to ½ nautical mile directly south of Colville Island. The rectangle's straight, neat edges lent a false sense of precision to the image. It would be my first tepid dip into the bottomless well of "believing what I wanted to believe." Chris's account did, however, have a strong supporting factor in its favor—he had witnessed 510 as it sank, or, at least, the jet's bubbles as it went under. Denby Starling only glimpsed the A-6 shortly before water impact, forcing him to extrapolate the approximate crash location.

The *Salvor* report curiously dismissed all unnamed eyewitness accounts of the accident, deeming the information unreliable. The report's gridded out, bare-bones map of the search area, however, still displayed a lone, hand-written label of "eyewitness." This witness's impact bearing was indicated by a single hand-drawn line of sight annotated "300" with no elaboration. Why was this onlooker's account included in the final *Salvor* report?

I started to view the *Salvor* report with skepticism. The reader was left to guess why certain evidence provided by an unnamed source was considered reliable while other critical data, such as eyewitness testimony, was apparently rejected wholesale. Exhaling in frustration, I walked out to the car for the ten-minute drive to the public library. Oak Harbor, a rural town

Bombardier-navigator Lt Chris Eagle (left) and pilot Cdr "Denby" Starling of VA-145 ejected safely from their A-6E this week and were whisked to awaiting medical care on shore at Naval Hospital. A quick response, the product of long hours of training, is credited with averting what could have been a tragedy. The men received only minor injuries. See story. (Lt. Paul Rogers photos)

Tragedy averted as 2 are rescued in waters off air station

An A-6E Intruder aircraft from VA-145 experienced mechanical difficulty while on a local routine training flight on Monday, Nov. 6. The aircraft impacted the water at about seven miles northwest of the naval air station.

Pilot Cdr "Denby" Starling, 36, and bombardier-navigator, Lt Chris Eagle, 26, both ejected safely.

The Navy Search and Rescue helicopter, piloted by LCdr Mike Ross and LCdr Paul Winberry, was over the scene in less than 15 minutes. Aircrewman AMSC Jimmy Lee Helton, Search and Rescue swimmer, went into the water from the helicopter to aid the aircrewmen as they were pulled to safety.

The crewmen were immediately flown to Naval Hospital at the air station, where they were treated for minor injuries and held for observation overnight. They were released Tuesday morning in good condition.

RAdm Grady Jackson complimented the rescue crew on the professionalism they showed in the rescue, and on a job well done.

At press time, no decision had been made concerning possible recovery of the aircraft.

Lieutenant Chris Eagle (left) and Commander Denby Starling (right) (*Crosswinds* newspaper photo, from the author's collection).

of about 20,000 people, exists at its current size only due to its primary employer, the U.S. Navy. The library system is a part of the greater, combined resources of two counties, and the facility located in Oak Harbor is a one-story building that shares space with a local community college. I walked in the front door.

The research librarian's desk sat like an island in the middle of the open room. I asked to see the archived *Whidbey-News Times* editions from 1989 through 1990, and the trim, brunette guided me to a microfiche viewer, the likes of which I hadn't seen, much less used, in thirty years. After a short tutorial, the librarian disappeared to retrieve the requested microfiche. She returned moments later with three small boxes of tightly wound film. I identified the box from November 1989 and started feeding the reel of two-inch plastic film through the machine's rollers. Fortunately, it was still early enough in the day that my meds were "on," allowing my fingers to negotiate the microfiche feeder maze reasonably deftly. The librarian left to help another patron.

The *Whidbey News-Times* was the area's primary civilian newspaper in 1989. It was also the only periodical at the library that contained anything at all about the mishap. There was a total of five articles written about the accident, with the opener running on the front page two days after the ejection, the first printing of the bi-weekly newspaper since the crash. The article was titled "A-6 bomber ditches near Lopez, fliers eject safely." Ditching is an unambiguous Navy term that, when used correctly, indicates the crew crash landed on the water. Ditching and ejection are mutually exclusive, which didn't help the piece's credibility. I kept reading.

One of the first sentences jumped out at me: "Starling, 36, is the executive officer of the A-6 squadron." I'd just turned 52. At the time of the ejection, the "old man" had been 16 years younger than my current age. I was a kid back then, I considered; we were all kids.

The *Whidbey News-Times* article continued with, "The crash occurred 7 miles west of N.A.S. Whidbey in the water near Lopez Island believed to be 200 to 250 feet deep."

The estimated ejection point identified in the *Salvor* report was 6 miles from the N.A.S. Whidbey TACAN, or "Tactical Air Navigation," a navigational aid used by pilots. A TACAN system included a cockpit gauge that showed the range and bearing to a transmitter/receiver unit either ashore or on a ship. The article indicated the crash occurred at 7 miles, 1 mile beyond the ejection point. For the Intruder to fly in a relatively straight line for 1 mile after the crew ejected seemed reasonable.

I spooled the microfiche forward to the next page. "Ejection is a very traumatic event. They were pretty lucky," a Navy spokesman said. "In the A-6, the seat blasts through the canopy because it cuts down on time spent waiting to eject. Of those who eject and live, the most common injury suffered is spinal. To prepare for ejection, aviators enroll in a two-day refresher course every four years to review ejection techniques and survival. A common fact taught in the course is that ejection crushes the human body with force 15 to 24 times greater than normal gravitational force. The human body is not structurally designed to withstand more than 18 times, or 18 Gs, the normal gravitational force. During a routine flight in an A-6, aviators usually experience a maximum of 4 to 6 Gs."

Except for the references to ditching, the article was technically accurate. Puget Sound was deep: a 200 to 250-foot depth could have described just about anywhere. Only a tiny demographic sliver of 1980s recreational divers ventured beyond 200-feet deep. At 200 feet, Puget Sound was 42 degrees Fahrenheit, or 5.5 degrees Celsius, and pitch black, allowing negligible visibility even with the brightest of underwater lights. The conditions were not all that different from my pre-Navy experiences wreck diving on the East Coast.

To explore these depths is to risk decompression sickness, the "bends," if a diver comes up too fast. The pressure of greater depth pushes any inert gas, such as nitrogen or helium, in the diver's breathing medium into liquid form in the bloodstream for dispersal throughout the body. To avoid the bends, a gradual ascent with stops at a variety of depths is required to vent off the expanding gas before it bubbles. Internal bubbling causes intense pain and damage to the body.

The bends aren't deep diving's only danger. Breathing regular air—nitrogen, oxygen, and a variety of trace elements—underwater can trigger a nitrogen narcosis induced state of diminished mental capacity. Nitrogen narcosis is like diving drunk, and it can be deadly. A deep diver also risks encountering oxygen toxicity, a little-understood physiological phenomenon that can abruptly send a diver into convulsions, usually leading to drowning. A final deadly hazard is hypercapnia, the result of breathing too rapidly at depth. This can lead to a dangerous build-up of carbon dioxide, hyperventilation, unconsciousness, and death.

Regular air, compressed and stuffed into high-pressure dive tanks, was the only gas used during all phases of a typical 1980s deep dive. In 1989, the word "trimix" had just been introduced to the recreational diving lexicon. Trimix replaces a portion of the nitrogen in a dive tank's compressed air with helium, a non-narcotic inert gas. Trimix was used in 2014 instead of air by most recreational deep divers. Consequently, today's technical divers can explore great depths without the risk of potentially debilitating nitrogen narcosis. The other dangers, however, still exist.

I printed out a copy of the first article before scrolling through the microfiche pages until finding the next reference to the incident. The December 6th, back page headline simply stated: "Navy might try to raise jet from ocean depths."

I read on. "The Navy is sending divers to explore the possibility of retrieving an A-6 bomber that crashed in the Strait of Juan de Fuca near Lopez Island November 6. Divers from Keyport Naval Undersea Warfare Engineering Station near Bremerton will dive several times in the next few weeks to locate the aircraft and determine whether a safe retrieval of the $30 million plane is possible. If the divers think a safe retrieval is feasible, the removal will probably be in late January or early February." It probably would have surprised everyone at the time to learn that 510 was no closer to being brought to the surface 25 years later.

I continued to the next page and did a double take. The next article was titled, "*Andrea Doria* artifacts display at Sno-Isle library." I sat up straight and shook my head. The coincidence was beyond strange; I had forgotten that this article existed.

The first line read, "(Navy) Lieutenant Pete Hunt was one of 16 divers who fought off sharks to surface from the depths of waters near Nantucket Island with the first 'publically disclosed chunk of treasure' from the sunken Italian ocean liner Andrea Doria in 1983."

I recognized the sentence as a paraphrase from the original *Long Island Newsday* article. In the Fall of 1989, a friend had arranged the library display of my *Andrea Doria* artifacts. I had not seen this story in 25 years and had no memory of it being connected in time to the crash of 510. It was bizarre to see a reference to my diving of thirty years earlier alongside the article about the missing A-6. I viewed it as a positive omen.

The next pertinent article did not appear for over two months, its headline erroneously proclaiming, "Navy begins to salvage jet from depths off Smith Island." The implication that the aircraft had been found was incorrect. The four-paragraph article briefly explained that the *Salvor* had arrived from Hawaii to join the search. Then my eyes froze on a sentence:

"The jet crashed about 6 miles west/northwest of N.A.S. Whidbey, about halfway between Lopez and Smith Islands." This was new. Why had the newspaper's reported crash site changed from 7 miles to 6? I worked my way to the end of the microfiche spool.

The fourth article about the A-6 crash was written two weeks later. One short piece, the last in the paper's five-part series, appeared a week after that. Both articles described the location of the crash site as 6 miles west/northwest of N.A.S. Whidbey. The last three stories had been written after the *Salvor* arrived in local waters. Did the *Salvor's* Captain change the assumed crash distance?

It was early afternoon when I left the library to go home for lunch and to review the *Salvor* report yet again. The break from my desk allowed for a fresh read of the report, and I noticed almost immediately that there seemed to be some confusion in the account between magnetic north and true north. Magnetic north is based on the actual location of the North Pole within the earth, which is somewhere in Canada. A compass needle always points to magnetic north. True north is a calculated direction that provides a consistent reference; it is always at the very top of a globe or straight up on most maps. The difference between magnetic and true headings is due to magnetic variation which diverges at higher latitudes. In Rosario Strait, the divergence between magnetic and true north—the magnetic variation—is 20 degrees. A heading of 300-degrees true north equals 280-degrees magnetic north in Rosario Strait.

The Lost Intruder

All headings and bearings listed in the *Salvor* report suddenly became suspect. Pilots do not use true headings. Flying with true headings requires constant reference to navigation charts to determine the accurate magnetic variation as it changes with location. For the *Salvor* report to list the lost Intruder's heading in terms of true north either meant that they had made the adjustment from magnetic north, or that they had made a mistake. The difference between 300 degrees in true versus magnetic headings in this instance was several miles, a potentially huge disparity.

It seemed that the Navy had started out by investigating random spots based on specific pieces of evidence, perhaps hoping for a quick find. When nothing was found, a more methodical approach was adopted. They then used the 300-degree radial at 6 nautical miles to draw expanding concentric circles, which were subsequently transposed into rectangular grid patterns to be searched. And they still couldn't find the missing A-6.

I took stock of the available information. Five-ten's crew probably ejected between 6 and 7 nautical miles west/northwest of the Naval Air Station. Chris Eagle's statement placed 510's water impact point ¼ to ½ nautical mile south of Lopez Island. His feeling now, 25 years after writing the statement, was that the jet hit the water ¼ to ½ nautical miles south of Colville Island. The *Whidbey-News Times* stated that the crash occurred between Lopez Island and Smith Island. Denby Starling's memory of the impact point was several miles south of Lopez, which better fit the newspaper's account.

It was difficult visualizing the geometry of the different information, particularly the distances, so I decided to plot all the data on a single chart. I booted up my laptop, opened Google Earth, and started marking the various points on the electronic map.

I flipped through the *Salvor* report until finding the initial latitude and longitude used as ground zero for the search. It was over a mile to the south of the *Salvor's* assumed ejection point. The geometry of the two locations made no sense. The report stated that the jet had been on a 300-degree heading, flying roughly northwest, when the crew ejected. It was extremely unlikely that after the ejection, 510 turned almost 180 degrees—half a circle—and then continued to fly straight ahead for well over a mile before crashing.

Furthermore, the 6-mile ejection point was almost 2 miles from either Lopez Island or Colville Island. If, as the simulator test trial indicated, the furthest the A-6 could have traveled was one nautical mile, then how could Chris Eagle's statement be correct? There was no reason given for the *Salvor's* 6-mile ejection distance from N.A.S. Whidbey. I began to believe that the *Salvor* had orchestrated their search effort a nautical mile short of where the crew actually ejected.

The difference between a 6 and a 7-nautical mile ejection distance from N.A.S. Whidbey Island was critical. The unnamed source of the TACAN fix—the only person with certain knowledge of the correct distance—was left unresolved. Standard operating procedures at the time, however, indicated that the most likely source of such information was the crew of 510's wingman, 502. Five-zero-two's bombardier/navigator would have been responsible for radio calls. That would make my old squadron-mate, Rivers Cleveland, the likely source of the TACAN information. It was still a guess, but at least I could ask Rivers directly and test his memory. I looked at my watch, added three hours for East Coast time, picked up the phone and dialed his home number.

I knew Rivers well. He had been my bombardier/navigator during our Operation Desert Storm deployment. We also flew together in Rivers' final flying billet, when he was the commanding officer of Attack Squadron 52. He now lives in Charleston, South Carolina, where he works for a defense contractor. Rivers picked up on the second ring, and we caught up briefly before I got to the point.

"Rivers, do you remember the source of 510's ejection range and bearing estimates? Was that your input?" I asked.

Rivers thought for a moment, then replied in a barely perceptible southern drawl, "Well, sort of. I gave air traffic control a general estimate of about 5 to 7 nautical miles west, northwest of the field. Why?"

Wonderful, I thought, now there was a third distance in the mix. "The newspaper accounts seem to change over time," I said, "None of the articles mention the source of the distances. Did somebody provide the *Salvor* a TACAN fix?" I waited.

The Lost Intruder

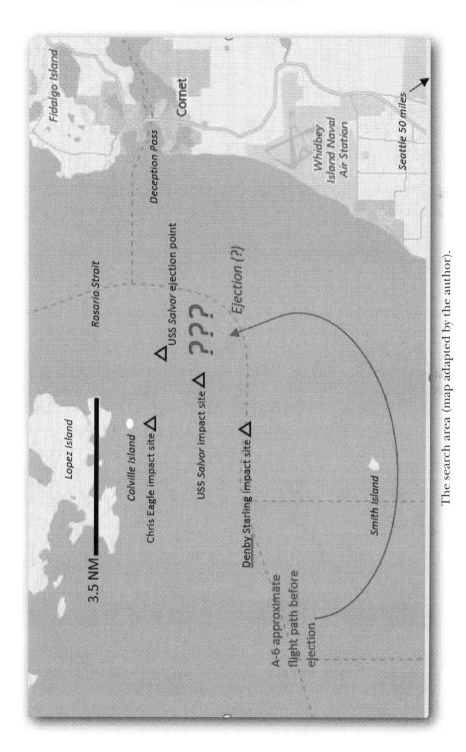

The search area (map adapted by the author).

"I suppose there could be a connection, but shoot, I wasn't referencing the TACAN anyway. I put the radar cursors on the threshold of the duty runway to give 510 a quick steer if things started to go bad. As it turned out, the situation deteriorated so quickly that I never had a chance to use the position. Five-ten was heading about 300 degrees when they punched out."

I said my goodbyes and hung up. It seemed that 5, 6 or even 7 nautical miles from N.A.S. Whidbey were all reasonable distance estimates for the ejection point. Fortunately, the jet's reported heading at ejection was consistently 300-degrees magnetic. The *Salvor* might have decided to split the difference with 6 nautical miles.

It was all theories about theories. The notion that the Navy had looked in the wrong area haunted me, but when I asked Chris Eagle about his decades-old opinion, he replied that he had no memory of ever saying such a thing. That didn't mean he never said it, Chris was quick to point out, he just couldn't tell me why he might have held that view so long ago. I didn't doubt my own memory; it made too much sense that the Navy looked in the wrong area.

If there was an unsearched portion of the Navy's survey grid, it was probably on the fringes, I reasoned, which made a 7-mile ejection distance more appealing. I returned to Google Earth and plotted 300 degrees from the tip of the active runway out to 7 miles. This moved the ejection point significantly further to the northwest. Suddenly, the evidence started to fall in line.

If the crew ejected 7 miles from the approach end of the runway heading 300 degrees magnetic, and then 510 continued to fly an additional mile before crashing, this would place the A-6 water impact point to the south, southwest of Colville Island. It would align perfectly with Chris Eagle's statement.

The sea floor in this coastal zone was rugged, with small islands and bays dotting the south shore of Lopez Island. It shot up from 300 feet to as shallow as 50 feet in less than half a mile. It would be tough to search this area, especially for Navy ships greater than 100 feet long. Going back to Google Earth, I drew a one nautical mile line from the newly revised ejection point and marked a ½ nautical mile buffer on either side. This seemed a reasonable distance for 510 to have deviated laterally. Next, I connected the dots and drew two large triangles joined by a shared border. This newly defined search zone was still a needle-in-a-haystack, but for the most part, it was a diveable haystack. We could spend years diving this area and never search it all, but at least there would be a minuscule chance of finding the jet. It was a working theory; one based on conjecture and sketchy assumptions, but a theory nonetheless.

In March, two friends from my old Navy days, Steve Kelley and Chris Burgess, chipped in to buy me a recreational quality Raymarine Dragonfly sonar/depth sounder as a birthday present. I now had a modicum of research materials: the *Salvor* report, Chris Eagle's statement, and interviews with three of the principals involved in the mishap. Add in the local newspaper coverage of the event, and I concluded that it was time to get my hands dirty by starting the physical search.

Still, a single question nagged at me more than the others. I desperately wanted to find 510 in shallow enough water to dive in my present condition. For that to be possible, it would have to be very close to shore. Was my theory a house of cards based on wishful thinking?

Given the rugged underwater terrain, the total surface area of sea bottom to be mapped was daunting. The recreational-grade Dragonfly sonar might have to pass directly overhead to find the missing A-6. Would it even look like an aircraft?

None of this bothered me. Not knowing the source of my confidence, I was sure of just one thing: somehow, I would find the lost Intruder.

Five

In painfully familiar direction

April 2014

Our first day on the water searching for 510 yielded less than impressive results. For three hours, a friend, my son, a soon to be seasick dog, and I motored slowly to the south of Colville Island, all the while staring in innocent anticipation at the Dragonfly's display screen. Although the experience moved us no closer to finding the missing jet, I did learn three important things: seasick dogs are a messy handful, the Dragonfly's operators manual deserved to be read, and the depth sounder's skinny sonar beam was like looking through a straw. Despite the dismal show of concrete results, actively engaging in the search—exploring—was still exciting.

The Dragonfly was clearly not a commercial grade piece of equipment. To start, it had no side-scan function. Side-scan sonar shoots energy pulses laterally hundreds of yards at a high angle to the vertical water column. This causes the smallest of objects sitting on the bottom to cast a shadow, magically becoming visible. The transducer is the sensor of a sonar system that physically emits the energy pulse and receives its return signal. Side-scan sonar transducers are housed in a torpedo shaped unit, usually several feet in length, that are towed just above the sea floor.

True side-scan sonar systems have three channels, or directions, in which they are able to gather information about the bottom. In addition to a side-scan's left and right side channels, which correspond to the left and right sides of the boat's track across the water, a down-scan channel is also available. The down-scan function is the least helpful of the three channels. This is due to its perspective of looking straight down, an angle that casts few shadows and usually reveals little sense of a contact's dimensions. But by making multiple, overlapping sonar runs from different directions, a side-scan sonar's three channels—left, right, and down-scan—can be "stitched together" by a computer program into a remarkably clear three-dimensional image of the underwater terrain.

The Dragonfly's lone sonar channel, trademarked "DownVision," produced a down-scan only perspective. The DownVision feature shot a sonar pulse straight down from the transducer mounted below the water line on *Sea Hunt's* transom, with an optimistically advertised sixty-degree field of view to either side of the boat. The image reflected from the bottom was displayed on the Dragonfly's modest 5.7-inch monitor.

This down-scan only setup had significant limitations. First, the DownVision sonar pulse would lose strength with each additional foot of water depth. For it to be effective anywhere close to its advertised limit of 600 feet deep, the bottom needed to be perfectly flat and the seas dead calm. Furthermore, a contact had to extend at least several feet up above the terrain for it to be seen. A boat exploring with a Dragonfly sonar had to be almost directly over a target if there was to be any chance of it being detected.

Scanning the search area's rugged bottom with the one-dimensional DownVision, I would soon learn, was a complete waste of time. The bottom was so covered with steep cliffs and jagged boulders that unidentifiable targets filled the tiny display screen. The Dragonfly did, however, still have a role to play in pinpointing a contact for future identification by divers, but only after first making an informed assessment of its composition using a fully capable side-scan sonar. Somehow, I needed to come up with one of the expensive side-scan units.

My search also required a clear-cut theory with sound assumptions, and there were two particularly troubling suppositions at my existing model's core. First, Chris Eagle's addition of Colville Island to the narrative after

25 years raised serious concerns. Chris recognized this when he characterized the use of Colville Island as the basis for his original distance reference as a "huge guess" in his reply email. There was also the questionable reliability of accurately judging distances while taking the first swing in a parachute, just seconds after violently breaking through the A-6 canopy glass. But Chris Eagle was an exceptional bombardier/navigator and an astute observer. My respect for his abilities made it easy to overlook the uncertainties of his eyewitness account.

Regardless of the factual grounding of the theory, I would almost certainly find contacts too far below the surface for me to dive. Not knowing where to start on the side-scan sonar issue, I decided to jump ahead and track down divers willing and able to help identify future deep-water contacts. Late one evening I texted a friend, Josh Smith, the slightly off-the-wall question, "How deep would you be willing to go to dive the wreck of an A-6 Intruder?"

I met Josh in February 2012 at the Northwest Dive Exposition in Tacoma. We started a friendship after he wrote a complimentary online review of my second book, *Setting the Hook*. Josh and his technical-diver friends were at the Tacoma Expo representing the Maritime Documentation Society (MDS), a shipwreck discovery and exploration non-profit started by my trimix and technical diving instructor of 2000-2001, Ron Akeson. I had lost touch with Ron over the years and didn't know much about the MDS.

Josh is a gregarious general contractor with a cleverly irreverent sense of humor that immediately had us hitting it off. Originally from New Mexico, Josh is married with no children. Although he lived several hours away in Seattle, we kept in touch, and he would occasionally stop by the marina to say hi. Still, it took me off guard when Josh immediately replied to my depth question with, "Three hundred, maybe four hundred feet." His enthusiasm was quickly matched by two of his MDS friends, Rob Wilson and Dan Warter.

Rob Wilson is married with a teenaged daughter. A private pilot and aircraft maintenance technician at Boeing Field in Seattle, Rob worked a stone's throw away from the A-6 I had dropped off at the Boeing Museum of

Flight in 1995. Rob, like me, had been diving since the 1970s. His thoughtful critiques of my theory proved particularly useful, especially during the first several months of the project. He also pressed several inconsistencies in my logic that I'd conveniently ignored in deference to maintaining a dive-friendly shallow water search area.

Dan Warter is a budding commercial underwater videographer who works in the dive equipment industry. He has made one successful documentary, *The Warrens: A Lake Crescent Mystery*, about the disappearance of a married couple in 1929, and was trying to stake his claim in the field with several more. He is married but has no children. Dan is a rail-thin, wiry-strong vegan, which made me feel old and out of place, but he, too, has a solidly irreverent sense of humor, which trumps just about everything as far as I'm concerned. Attracted by the allure of finding and diving a virgin military aircraft wreck, the three technical divers—Josh, Rob, and Dan—immediately offered to help.

Explaining my theory to others highlighted my desperate shortage of hard facts. One rare piece of available firm evidence was the exact timing of the ejection sequence relative to Chris Eagle's observation of 510's water impact. I had kept a copy of the dense A-6 NATOPS manual—the "*Naval Air Training and Operating Procedures Standardization*" bible of Intruder aircraft systems and procedures—after leaving the Navy. I opened the NATOPS to its detailed ejection schematic and reread Chris Eagle's statement.

"I grabbed the lower ejection handle, leaned back in the seat, and pulled. I recall the initial kick of leaving the cockpit and the sensation of my head being tossed about, but all was black at this point. The first thing I remember seeing was while I was still in the seat. I was looking down past my left leg and could see the aircraft below and slightly ahead of me as I brought my head upright."

The A-6 ejection mechanism was designed to keep the crewmember tightly strapped in the seat momentarily after ejection. This was to stabilize the ejected aviator in the wind stream and reduce the chances of flail injuries. Flail injuries could occur if a crewmember's arms or legs were caught in the wind stream during a high-speed ejection, usually resulting in nasty breaks. I read on.

"I saw what I believe to have been (Denby Starling) still in his ejection seat off to my left and slightly in front. I felt a tumbling sensation as the seat began to roll backward and shortly after that was jerked out of the ejection seat. The next thing I remember was hanging in my parachute shaking off some stars in my eyes. I looked down and saw a bubbling white circle in the water which I assumed to have been where the aircraft impacted."

If I could figure out how long it took for the ejection seat to automatically deploy its parachute, then I would be able to determine 510's approximate travel time from the ejection point to the crash site. All that would need to be added was the time it took to "shake off some stars in my eyes."

After scanning the notes in the NATOPS ejection schematic, I discovered that it took precisely 4.9 seconds from the pull of the ejection handle until the main parachute inflated. Assuming 510 continued flying straight ahead, the A-6's course line could be extended forward 4.9 seconds from where it was theorized that the ejection had occurred. Add in the time required to "shake off some stars," and we would have the distance to the water impact point in seconds. An aircraft's safe flying speed at landing varies with its gross weight, which is mostly determined by the amount of fuel it is carrying. Five-ten at the time of ejection was at its maximum weight landing speed, which translated to 120 knots indicated airspeed, or two nautical miles per minute, discounting the effects of any wind. Determining the distance traveled from an elapsed time was basic math once the approximate speed of the aircraft was known.

I explained my thought process to Rob Wilson in an email, and his reply abruptly threw a wrench into what I had thought was a well-considered theory.

Tuesday, April 1, 2014

Pete,

I'm getting a bit hung up on the 4.9-second ejection sequence. My understanding is 4.9 seconds from ejection initiation to parachute deployment, and the jet was reasonably stable during the ejection sequence (it was not screaming straight down). The jet impacted during this time. It just seems a real short amount of time to cover 6000 feet of altitude.

Rob

Rob had highlighted a big problem: 4.9 seconds was indeed just too little time. Traveling at 120 knots equals about .033 nautical miles per second. Multiplying .033 by 4.9 seconds yields a total of .16 nautical miles flown downrange before impacting the water. With a starting altitude of 6,000 feet, that would have required 510 to nose over at an exceptionally steep angle almost immediately, an almost impossibly steep dive angle. Even assuming a nominal ten seconds for Chris to shake "off the stars" from the parachute's opening shock, the numbers made no sense. No jet properly trimmed for 120 knots airspeed could accelerate and descend that quickly with flaps and slats down—6,000 feet was just too high above the water.

This uncomfortable quandary would bother me for months. My faith in Chris's opinion, bolstered by years of hearing him re-tell the story, was absolute. But what if Chris Eagle was wrong? What if he had been looking at something else making a "bubbling white circle of water"?

Later that week, I was out on the water again, this time, to travel to Lopez Island where my teenaged son Jared would participate in a regional sailing regatta. The trip offered an opportunity to investigate the San Juan Island area newspapers for A-6 related stories from 1989. After several phone calls and an online search, I learned that none of the local papers—and there were five of them—had a computer database going back to 1989. The Lopez Island Library did, however, have hard copies of the *Lopez Islander* going back that far. I walked the half mile country road from the marina to the Lopez Island Library to search the archived newspapers and found two related articles.

The first story indicated that, shortly after the accident, what was thought to be either an aluminum wing or fuselage component had been found at a park on nearby San Juan Island. The park was about four miles to the west of Colville Island. A passerby had retrieved the debris from the shore before walking it up a small hill to a picnic bench. I knew from previous research that the ejection had occurred about the same time as the high slack, the point where the tidal waters go still for a short period before ebbing. Realistically, the most that could be concluded was that the spot where the debris made landfall did not outright contradict my theory's water impact point.

The next reference to the A-6 debris came on December 6th, 1989. The gentleman who had found the aircraft part had come forward after reading the local article. He elaborated that he had no idea that the debris was from a jet: he assumed it to be a part of a broken-up Coleman cooler. He visited the park regularly and had been watching the debris floating in the surf for several days before fishing out a piece. That was the last reference to the crash in the *Lopez Islander*. The intervening tides for the several weeks between the accident and the recovery of the debris, I concluded in frustration, could have moved the airplane part just about anywhere.

Signs of encouragement, however, did keep popping up in the most unexpected ways. The week after the sailing regatta, I drove with my son the twenty minutes north to an orthodontist appointment in the town of Anacortes. Jared was called to the back of the office almost immediately as I shuffled into the waiting room to sit down. Looking up, I was surprised to recognize a familiar face. I first met Richard at a local gym about seven years earlier, and I remembered that he was a paramedic with the local Whidbey Island hospital's ambulance fleet. He had been in the Navy at some point in an earlier career. Acting on a hunch, I asked Richard what his job had been during his time in the service. It turned out that Richard had been a rescue swimmer. I asked the obvious question, "What was your assigned unit at N.A.S. Whidbey?"

He replied, "I was stationed at Whidbey from 1985 through 1991 as the Crew Chief for the search and rescue helicopter." My jaw dropped. The Crew Chief was the enlisted man in charge of the search and rescue—SAR—helicopter crew that plucked Denby Starling and Chris Eagle from the water. I couldn't believe the luck.

"Do you remember the A-6 that went down over the water in 1989?" I held my breath.

"Sure, I worked that rescue, remember it well." Seriously, what were the chances, I thought? In an orthodontist's office?

I asked if he remembered the exact location of where they had found the aircrew. Again, Richard replied without hesitation, his memory sound on this point.

"When we picked up the crew, I distinctly remember the helicopter being directly west of the Sound View Shopper." The convenience store was located just to the north of N.A.S. Whidbey Island.

I pressed Richard for more details, but that was all his memory allowed. Still, it was helpful. If enough eyewitness lines of bearing could be drawn, then these lines might converge, perhaps eventually intersecting at common geographic points. Richard's memory of the event might just provide an additional small clue. Jared finished with his appointment, and we drove home.

I spent the following several days communicating with the MDS members, fishing for ideas on what to do next. The three MDS divers were proving to be knowledgeable sounding boards, often raising questions I had not considered. This was not surprising; after all, it wasn't their first underwater search. Back in 2000, the MDS had been largely a dream of Ron Akeson's put on hold. I only became aware of the existence of the MDS precursor, the "Deep Green Dive Team," while searching the Internet for Ron's dive store, Adventures Down Under.

The Deep Green Diver Team didn't have much to show for itself in 2000. There were just two active members of the fledgling group, Ron and his friend John Campbell. In 2011, the MDS was officially formed, and priority was placed on increasing membership and finding new projects. By 2014, the MDS actively scheduled challenging technical wreck dives each season, sharing their discoveries at scuba diving expositions and other venues. Ron Akeson's dream of creating a credible organization from which to launch their explorations had become a reality.

And then, on April 18, 2014, tragedy struck. During a technical training dive at Mukilteo, a beach north of Seattle where Ron had made hundreds if not thousands of dives, something went horribly wrong. Ron and his two students were using rebreathers, relatively recent additions to technical diving's innovations. Modern rebreathers substantially increase the time a diver can spend on the bottom by recycling the oxygen breathed within a closed loop. A mere fraction of air's content of oxygen is consumed by the body when taking a breath.

By removing—or "scrubbing"—the carbon dioxide from the gas mixture, the amount of reclaimed oxygen can extend a dive by hours.

Additionally, the proportions of oxygen and trimix breathed during any phase of a dive can be fine-tuned automatically as desired. Varying the ratio of oxygen to trimix in a diver's breathing mixture allows for greater depths, less narcosis, shorter—but still extremely long—decompressions, and considerably more gas available to the diver in case of a problem. Assuming, that is, that the problem does not involve a catastrophic failure of the rebreather itself.

In the ultra-familiar waters of Mukilteo, at the unremarkable depth of ninety feet, Ron ran into trouble. It is unclear to this day what exactly occurred. When mysterious accidents happen to highly qualified people, such as Ron, the inevitable search for meaning can be depressingly futile. Something occurred underwater that kept Ron from breathing for just five minutes. It was long enough to destroy his brain.

Ron's death had a devastating impact on the MDS. Whether due to a desire to be active during difficult emotional times, a genuine interest in a unique search, or both, the key MDS players were looking for a project. The MDS officially voted to help with the hunt for the lost Intruder. I gladly accepted their offer.

Six

The turn of temptation approaches

May 2014

The lost Intruder project took a significant step forward on the morning of May 3rd as five men I barely knew showed up at the Deception Pass Marina to assist in a day of sonar scanning. Lacking self-confidence for the first time in my life, I struggled with a sense of vulnerability; a disease induced free-floating anxiety that others might try to take advantage of the situation in some inexplicable way. I fought hard to welcome the mostly strangers to the project.

Josh Smith is a down to earth big guy with a friendly air about him. He was also the personal contact who, along with Rob Wilson, a man I knew through Ron Akeson, instilled a long-term trust in the MDS. Four of the five guests on the boat that day were MDS members: Josh Smith, Dan Warter, Rob Wilson, and Ben Griner. John Rawlins, the fifth visitor aboard *Sea Hunt*, was not affiliated with the MDS but had been a reputable technical diver for many years.

I knew nothing about Ben Griner other than two critical facts—he owned a side-scan sonar and was an expert in its use. Even better, he seemed interested in finding the missing A-6. Oddly, none of this surprised me. I had an intuitive sense that things would somehow work out, and time and again, that is precisely what happened.

I explained my revised ejection-point theory to the group through a series of emails in the days before we all met at the marina. We left the dock at 6:30 am in a brightening sky and calm seas. Ben had carried aboard several large, hard sided plastic suitcases containing a side-scan "towfish" sonar with portable battery packs, hundreds of feet of spooled transducer cable, a laptop, and an extra-large, remote computer monitor.

John Rawlins and Ben Griner had recently stopped diving due to unrelated medical conditions, but Josh, Rob, and Dan all hoped to be in the first cadre of divers to visit the downed A-6. Dan Warter also let it be known that he was entertaining the idea of making a documentary film about the adventure. The way that he framed the proposal gave me veto authority if I saw a conflict with a future book. I appreciated his candor, but not wanting to pass up an opportunity to give him a hard time promptly told him in a slurry, Parkinson's mumble that "There's not a chance in hell that I would allow such a thing." Dan took this introduction to my off-key sense of humor surprisingly well, although he is such a nice guy that I'm not quite sure he knew what to make of me. Let's just say that I'm still waiting for the documentary.

Ben Griner is a solemn individual of medium height and build. Married with two children, Ben was tight-lipped about his family and most everything else in his personal life. I would end up spending scores of hours with Ben in *Sea Hunt's* cabin, far more than with any other team member in 2014. During much of that time, tensions ran high as we worked in challenging conditions of extreme current and adverse weather. From the beginning, Ben fully committed to the project with his time, effort, and equipment.

Ben and Dan were used to working together, having successfully discovered several World War Two vintage airplanes in Seattle's Lake Washington. Although I never seriously doubted that we would find 510, the addition of Ben and Dan to the team gave me a sense of how we would locate the missing jet. We crisscrossed the rocky bottom shallows to the south of Lopez Island for ten non-stop hours, pulling the 18-inch, cylindrical towfish hundreds of feet behind the boat. The survey was conducted painstakingly slowly as we overlapped each previous run by 200 feet laterally. The sidescan could collect imagery as far out as 800 feet to either side of *Sea Hunt's* course line or about a quarter of a nautical mile total—half to the left, half to the right—with each pass.

Ben and Dan stood hunched over the laptop's sonar display at the rear of the cabin, discounting each sonar target with a tersely exclaimed best-guess description of dismissal, such as "rock pile" or "sunken log." When pulled at the appropriate speed, the towfish would drop down close to the bottom, at times barely a dozen feet from the jagged boulders. Ben judiciously took notes on every sonar contact, knowing that someday the data might come in handy. But scarce few images the size of an A-6 Intruder crossed the laptop display, and almost none that looked even remotely like an aircraft.

The waters south of Lopez Island are fertile fishing grounds, and dozens of recreational fishermen were out to take advantage of the bright, spring day. Due to *Sea Hunt*'s limited maneuverability trailing the sonar cable, it proved inevitable that our course would eventually conflict with that of another vessel. When this happened, I took a cue from Ben and waved to the approaching boat out the window while pointing to the sonar cable strung out behind us. All of the vessels forced out of *Sea Hunt's* slow and steady path gave way graciously. Such is the advantage of working around recreational boaters, all of whom are out on the water by choice to have fun. The few larger commercial vessels transiting the shallow coastal waters were almost exclusively from the whale watching fleet. The whale watching captains listened to their radios diligently, however, and given sufficient time to respond would usually take a slight detour around *Sea Hunt* without complaint.

We had yet to enter the shipping lanes, several hundred yards to the east of our inwardly spiraling search pattern, where such courtesies would not exist. The tankers transiting the Straits reached nearly 900 feet in length, barely allowing sufficient room to maneuver in the narrow, deep-water section of the passage. If a tanker Captain were willing to turn a few degrees off course to avoid our path, then the coming conflict with *Sea Hunt's* course would need to be recognized many miles away. Otherwise, there would be insufficient time for the ship's turn to take effect. The plan had *Sea Hunt* venturing into the shipping lanes only if it became necessary to search waters deeper than 200 feet.

My ability to tolerate the long work day initially came as a surprise. Although I didn't consider it at the time, my new-found resilience was the result of several factors. First, the area we were surveying was relatively easy to maneuver within. The clear sky and proximity to shore provided intuitive reference points, making navigation easy. There was no need for the helmsmen to engage in the tiring task of staring at the dully glowing navigation monitors for hours on end. Once on a steady course, the helmsman could look out the window straight ahead, choosing among myriad island landmarks to align in a navigation bearing. This made the endless chore of steadying *Sea Hunt's* light bow against the wind and current far easier than doing so using just the boat's navigation instruments. Having a professional grade side-scan sonar with an experienced operator along boosted my morale and confidence as well.

The division of responsibilities during a sonar search was relatively straight forward. The helmsman would reference the remote computer monitor, placed for easy viewing on the small galley counter next to the helm. The helmsman would then steer along the contoured edge of *Sea Hunt's* prior run, thereby automatically creating the desired overlap in sonar coverage. If the waters started to shallow, Ben would tell Dan Warter, or whoever was the deck hand for the day, to haul in some of the cable at the back of the boat. This would allow *Sea Hunt* to maintain a steady speed without causing the cable to drop excessively, possibly dragging the expensive sensor along the rocky bottom. In deeper waters, we would eliminate too much cable sag by increasing the boat's speed, like running faster with a kite to make it go higher.

The tricky part was maintaining a constant speed over the water in conditions of constantly shifting currents and wind. The local tidal exchange regularly exceeded three knots and could change abruptly in direction and magnitude. To maintain a three-knot speed over ground going directly into a three-knot current required substantial power. Conversely, a fast current pushing on the stern would cause the boat's speed of advance to immediately climb more than three knots, too fast for the towfish to obtain meaningful data. Complicating matters further, if power were retarded to counter a current's push in an attempt to slow, all headway—the boat's directional control—would be lost. *Sea Hunt* would then wallow with wide swings of the bow at the whim of the current and wind, allowing the boat to be pushed sideways, or even rotated backward, all the while heading off course.

In a routine that would have been comical if not for the potentially disastrous financial risk of losing the towfish, each turn reversal introduced wild steering oscillations and often less than smooth power corrections. I would attempt to impossibly maintain three knots over the ground with a two to three-knot current pushing the stern. Ben would test his patience with my apparent incompetence. Each sonar lap might extend a mile long, allowing time for the current to shift on *Sea Hunt's* return track. This could make what was impossible on a previous run just half an hour earlier surprisingly easy on a repeat run, and vice versa.

It took me a long time to accurately analyze the effects of current and the wind on the boat and the towfish. Almost always playing catch up, I was quick to blame my erratic boat-handling skills as the entirety of the problem. This was, in part, because I was grateful that Ben was donating his time, never mind chancing the loss of the towfish, in what was essentially a favor to me. With each lap around the track slightly different from the others, it was hard to determine the current's actual impact on each side-scan run.

At first, I accepted total responsibility for not being able to control the boat. It could be tricky driving, and some of the mistakes—20% or 30%, it was impossible for me to tell—were indeed my fault. Ben, who was most recently accustomed to surveying a calm, current-free lake, tried to understand why I had such difficulty maneuvering *Sea Hunt*, while I lacked the ability to accurately describe the problem. My situational awareness in recognizing when conditions became untenable was virtually non-existent.

Peter Hunt (left) and Ben Griner (right, sitting) taking a break on *Sea Hunt*. Dragonfly sonar 5.7" display is behind and to the right of Peter (from the author's collection).

As the sun lowered into late afternoon, a surprise wind storm blew up from the south without warning. Within minutes, the seas were too rough for the towfish to paint a reliable picture of the bottom, and it was all I could do to point *Sea Hunt's* bow into the wind to avoid broaching in the five-foot swells. The winds finally shifted just enough to allow for a bee line run to Deception Pass, but it took twice the normal time to return to the marina.

I fought to keep my eyes open as *Sea Hunt* approached the swift maelstrom of currents under the Deception Pass Bridge. As the sole passage of water around the north of Whidbey Island, massive volumes raced through Deception Pass during each of the four daily tidal exchanges, creating the fastest currents in Washington State. At times, the water would literally stack up, unable to transit the narrow Pass as quickly as the moon's force dictated, occasionally creating a difference in water height clearly visible to the naked eye. The unexpended energy of the frustrated water volume would manifest itself through whirlpools, back-eddies, and unpredictable deflections off the rugged bottom terrain and jagged cliff faces.

The Lost Intruder

A familiar anxiety grew in my gut like an ulcer at the prospect of nodding off at a critical juncture. Cycling hands back and forth between the radar and chart plotter, changing ranges, adjusting the picture, I tried to focus the disease commanded movements in a productive direction. The incessant beeping of the autopilot with each degree of heading change was as irritatingly useful as an alarm clock, rousing me back from the darkness of fluttering eyelids again and again. Finally, through the worst of the eight-knot current, we made a horseshoe turn to the west around Ben Ure Island and pointed toward the marina channel. Once on calm waters, I began to feel better. My choppy exhalations disappeared. Suddenly starved for oxygen, I took a long, deep breath, relaxing my muscles for the first time in hours.

After parking the boat, I watched the five passengers unload their gear. We had covered a vast tract of underwater real estate. Even better, before leaving the shallows of Lopez Island, Ben had made it clear that at least one contact held promise. It would be an incredible stroke of luck to find 510 on the first day of the side-scan search, I thought, only to be overcome moments later with a surge of confidence that we had indeed found the missing jet. This challenging first full day of searching made me question whether the combination of Parkinson's and fatigue was making me subtly lose touch with reality. There can be an element of mild psychosis to Parkinson's, and despite my protestations of unaffected cognition, I struggled, and continue to struggle to this day, with an accurate answer to the question.

Before Ben left, I requested a quick debrief. After gathering the rest of the crew, Ben systematically laid out a plan for what came next. It would take him about twenty hours of computer time to sort out the raw data. His specialized software could then optimize the imagery by building a three-dimensional mosaic of the promising contact. Frequently glancing at Dan Warter for support, he spoke about the problematic nature of finding submerged aircraft wrecks in general. Each of the Navy airplanes discovered by Ben and Dan in Lake Washington had presented unique identification challenges. In the end, it always came down to getting eyes on the target, be it either a diver's or through the use of an ROV.

Ben was confident that the day's sonar contact was as likely to be a Navy aircraft as any of their Lake Washington aircraft finds. Even better, it lay in just 145 feet of water. The seed was planted for me to dive the contact from the moment I heard its depth. The rest of the crew left for home while I straightened up the boat.

For the next few days, Ben worked on the project imagery in his spare time while I continued with the research. Deciding to take a new tact, I searched the Internet for information on any other aircraft that might have crashed near Rosario Strait. Our A-6 was not alone. Almost immediately, I came across a 150-page academic paper written in 1996 entitled "*US Navy Shipwrecks and Submerged Naval Aircraft in Washington: An Overview.*" The report's authors were members of an organization called the International Archaeological Research Institute. Although there were only 5 Navy aircraft wrecks within a 15-mile radius of Rosario Strait, it listed an astounding 44 total Navy aircraft losses in Washington State. Most of the crashes occurred during training in World War II, with 8 of the downed aircraft near Sand Point Naval Air Station along the shores of Lake Washington. At least 2 planes had crashed in the same general area as 510.

Both aircraft—a TBY Avenger and a Grumman Hellcat—were World War II vintage propeller planes. The airplanes were 40 and 33 feet long respectively compared to the A-6's length of 55 feet, but the wingspans of all three aircraft were within 15 feet of one another. Just how much of each wreck's fuselage might still be intact after so many decades in the corrosive salt water was anyone's guess, but perhaps enough to possibly introduce confusion if we ran across the wrong aircraft.

The Avenger ditched due to a mechanical malfunction in the southern portion of Rosario Strait. The crew survived the crash landing unscathed, leaving them to watch their intact Avenger sink, presumably from the safety of a life raft. The Hellcat crashed about five miles northwest of Smith Island after a mid-air collision. It probably broke apart airborne. The Hellcat pilot bailed out, but he was never found and presumably drowned. It was exciting to think that there was not just one, but three Navy aircraft somewhere below *Sea Hunt's* keel and, thanks to Ben and his sonar, we had a real chance of finding them.

The Lost Intruder

It didn't take long before Ben sent an email with good news.
Tuesday, May 6, 2014
Dan and Pete,

I have begun processing the (sonar) data sets. Dan and I noted a target that looked initially like a tree with a single branch. On analysis, we do need to look at this target either with divers or an ROV. After re-reading the research sent by Pete, I feel we really need to clear the remaining shallow areas. So I wanted to start the discussion around scheduling another trip out. How does Saturday look for each of you?

Ben

What motivated people like Ben? What drove him to risk the loss of thousands of dollars of equipment, spending countless hours helping out a person he had just met? The lost Intruder project certainly offered a unique challenge—finding the jet would accomplish something that the U.S. Navy had failed to do just weeks after the crash. And we might finally discover why the Navy search had been unsuccessful. Still, it did not fully explain Ben's level of motivation and sacrifice. On our next search outing, I asked Ben directly what he hoped to gain from the project.

He responded without hesitation, "I see the effort and commitment you're putting in. Most people give up after the first day of looking when it becomes apparent that no results are guaranteed and that it might take a very long time to fail. I love to explore, and it's not often that I meet someone with the persistence to stick with it to the end. So I'm in."

I listened, impressed, and wondered what would define the end of the project if 510 was too broken up to be found. Would the impossibility of success even be recognizable? Could persistence alone possibly be enough?

Seven

...OBSCURE MISDIRECTION OR DOUBT

Memories of 1989

The lost Intruder had no combat record or any particularly historic mission. The Navy only initiated the hunt for the missing jet to discover the cause of 510's total hydraulic systems failure, the mechanical malfunction that ultimately doomed the aircraft to the depths. Five-ten's hydraulic anomaly was not the first such occurrence in the Intruder fleet, which made it all the more disturbing. Five-ten was, however, the first A-6 to encounter this particular problem, crash, and then sink in sufficiently shallow water to be potentially salvageable. The only other similar mishaps had occurred far out at sea, in thousands of feet of open ocean.

A total hydraulic failure in an A-6 with no external cause—like a bullet—was not supposed to happen. It pointed to a design flaw, making it a near certainty that it would eventually happen again. The mystery of why the main landing gear failed to extend added urgency to the Navy's search. The price tag of an Intruder, never mind the danger to future aircrew, made finding 510 a priority.

Studying Chris Eagle's written statement brought back vivid snippets of memory: the faintly rotten smell of the freshly waxed ready room floor, the crackle of the base radio on the Squadron Duty Officer's desk. I could almost hear the background hum of the ready room, replete with the good-natured banter of coffee drinking junior officers. My reminiscence shifted to a rustic image, a modern-day Norman Rockwell style snapshot of Americana: a handful of young men in green, loosely hanging flight suits, beyond the superficiality of the uniforms and haircut, each subtly distinctive in appearance. In memory's vision, all shared a barely perceptible common purpose in their eyes, standing at ease under the fluorescent lights in the pre-dawn darkness. The anticipation would have been palatable, with each man cautiously hiding his eagerness for the first hint of jet noise, the unambiguous signal marking the commencement of flight operations. Researching the lost Intruder inspired many such moments, as the past was brought to life in my mind's eye to the smallest detail. I was reminded of long forgotten habits, an entire way of life, in so many ways completely foreign to me now.

I could remember procedures and checklists from decades earlier, but was calloused to the era, even when considering friends lost and the unseen, de-personalized enemies killed. It was another life from which I was firmly separated, those fast-moving years when 510 soared mightily through the skies. For fourteen years, 510's engines marked with sharply defined contrails her circular progression from somewhere to nowhere, and now—maybe—back to the world of human interaction. My connection to the lost Intruder was only remarkable for its lack of emotional significance. How, then, to explain my profound commitment to finding her?

I could only rationally define my persistence through what the lost Intruder did not mean to me. The A-6 at the bottom of Rosario Strait did not offer a return to youth, or even a glimpse of who I had been as a young man. All it promised was proof that it was not just the memories of self-delusion. Finding 510 would provide actual physical evidence of my life 25 years earlier. It offered a sliver of stability. In a world of shaky life assumptions and constant physical change, it proved to me that I was still sane.

The level of detail in Chris Eagle's antiseptic statement was not surprising. Tall with light skin, faintly red hair, and wiry thin, Chris did not suffer fools gladly. He looked a bit like Ron Howard, closer to the timeframe of "Opie" of the *Andy Griffith Show* than of the famed director of today. He was a tireless worker and talented radar operator with deadly bombing skills, perhaps the best bombardier/navigator in the Navy. I trusted his opinion implicitly. According to his statement, there had been an All Officers Meeting that Monday morning, a typical occurrence to start the week. It would have been staged in the ready room, on the second floor of the large aircraft hangar that also housed the squadron's administrative and maintenance spaces.

All Officers Meetings were an opportunity for the Skipper to set straight any issues in officer performance, be it lackluster preparation to fly or insufficient attention to collateral duties. Praise, if warranted, was given out as well. The meetings started at 8:00 am sharp. Typically, the gatherings would feature a ten-minute aircraft system refresher brief conducted by a squadron junior officer, followed by a short quiz. It was normal for each squadron to have a few skittish junior officers pouring over densely written NATOPS manual pages late into Sunday evening, trying to salvage a nugget of interest from the impenetrable tome. According to Chris Eagle's statement, the squadron's assigned briefing topic on the overcast morning of Monday, November 6, 1989, had been—as fate would have it—the A-6 hydraulic systems.

Upon completion of the All Officers Meeting, Denby Starling and Chris Eagle went to their respective offices. It was easy to picture the mostly empty halls, with the occasional sailor or aviator venturing out just long enough to dart into a nearby doorway to continue a slow, but steady, pace of work. The barely audible hum of serious conversation would gradually increase until the squadron woke up from the Monday morning doldrums and prepared to fly.

It is unknown whether Lieutenant Eagle had any pressing work to do that Monday morning. Commander Starling, on the other hand, almost certainly did. The squadron's second in command was always loaded down with stacks of paperwork. He was also responsible for unit discipline and the dispensing of justice if required. Denby Starling filled both roles exceedingly well. His boyish looks were overshadowed by a perpetually serious demeanor: the life of a squadron Executive Officer was hard work, and Starling took it every bit as seriously as he did flying. Starling was tough but fair, a rare mid-grade officer in that he was both liked and highly respected.

The Lost Intruder

I doubt that it surprised anyone decades later to learn that Starling had retired from the Navy as a senior admiral—he deserved it. More to the point, Starling had served the Navy, and the country, exceedingly well. The coincidence of the sharing of the men's last names with different types of birds had escaped me until starting my research. The easy going nature of the observation just didn't square with the relative seriousness, even gravitas, of both "Starling" and "Eagle."

The cloud layers were spaced too closely together on this Monday morning for the flight's scheduled defensive air-to-air combat training. Commander Starling instructed the duty officer to check for the availability of a low-level route time slot, and the mission was changed to a two-plane formation low-level flight to the target range in Boardman, Oregon. Maintenance was directed to upload six Mk-76 practice bombs armed with daytime smoke charges onto each Intruder.

Commander Starling decided that the second A-6, nose number 502, crewed by pilot Lieutenant Ray Roberts and bombardier/navigator Lieutenant Commander Rivers Cleveland, would assume the lead position. Ray Roberts would brief and direct the conduct of the two-plane flight. This way, the sortie would count as a syllabus training hop for Ray on his path to earning a qualification as a section lead. A section flight was made up of two jets; a division consisted of four. Commander Starling would fly in the wingman position to evaluate Lieutenant Roberts. This was not the usual position for a senior squadron officer to fly, but it was the accepted way of allowing younger aviators to gain the essential leadership experience necessary to earn higher qualifications.

I visualized the four airmen gathered at the front of the ready room. All except Ray Roberts would have been sitting on a piece of desk-chair combination furniture, identical to the type found in a high school classroom. Ray would have been standing, ready to brief the other three men. As a regularly assigned crew, Starling and Eagle would have sat together, while Rivers Cleveland might have been perhaps ten feet away, leaving Roberts an empty chair for when he finished the group instructions. Ray started precisely on time with three loudly spoken words that shut down all other ready room conversation:

"Attention to brief."

Once the briefing was complete, the four men stowed low-level charts and briefing guides into navigation bags to take with them to their respective jets. Each paired crew then walked side-by-side down the stairs to the parachute loft at the edge of the hanger bay to don their flight gear. After pulling on tightly fitting anti-G suits, leg restraints, and inflatable survival vests, each grabbed their helmet and oxygen mask before working their way to maintenance control to review their aircraft's log book. Five-ten's log book was thick with individual sheets of three-hole punched green paper. The old maintenance action forms described servicing completed, repairs made, and problems related by previous aircrews that could not be duplicated.

Chris Eagle's statement mentioned a single outstanding mechanical issue in 510's logbook, an electronic component on the bombardier/navigator's switch-laden armament panel. The armament panel sent the electrical signals to drop—or fire, in the case of a missile—ordnance in different quantities and types of attacks. An armament panel problem would not have affected 510's hydraulic systems or landing gear. Starling and Eagle walked out of maintenance control and then crossed the tarmac past the line of Attack Squadron 145 Intruders, each painted a dull gray to blend in with the ocean. The two men split up after reaching the A-6 with "510" written above the engine cowling, each going to opposite sides of the jet to complete an exterior inspection.

I knew these details to be accurate because it was always done this way. Training demanded consistency in habit patterns. The lives of the crew literally depended on the disciplined standardization of regular procedure.

Ray Roberts and Rivers Cleveland were doing the same thing in 502 elsewhere on the flight line. The pilots each started their respective jet's engines, then taxied independently until the two Intruders met up in the radar warm up area, a designated taxiway where A-6s could safely test their powerful ground mapping radars. When ready, Commander Starling would have given Ray in the lead A-6 a thumbs up, indicating a readiness to taxi together as a section to the active runway. Except for the target runs at Boardman, 502 and 510 would maintain formation, flying as lead and wing for every phase of the mission until it was time to land.

The Lost Intruder

Approaching the duty runway, the control tower cleared the flight to take "position and hold." The pair of Intruders taxied onto the runway and completed their final takeoff checks, identical except for one item: the position of the transponder switch. Once airborne, a transponder highlighted an aircraft's radar position and an altitude read-out to air traffic controllers. Only 502's crew, in the lead position, would have turned their transponder from "standby" to "on." If both aircraft transponders were activated in a formation flight, the closely spaced attenuated radar returns could be misconstrued by air traffic control as an impending mid-air collision. It was standard procedure, however, for the wingman to pre-load the flight's assigned transponder code for immediate use in the event of an equipment malfunction in the lead Intruder. Barring a problem, the wingman would leave his transponder unpowered in the "standby" mode for the duration of the flight. Both crews dialed the number 4667 into their respective aircraft's transponder.

Ray Roberts centered 502 on the left half of the runway as the section prepared for a formation takeoff. Five-ten had just gotten into position on the right half of the runway when the tower cleared them to go. The deafening roar of two sets of Pratt and Whitney J52 jet engines going to full power could be heard for miles as each A-6 shuddered and strained against their brakes. Rivers Cleveland raised his right hand above 502's canopy bow, looked for a nod of ready from Starling, and with a decisive, downward motion gave the signal to release brakes. The pair of 50,000-pound Intruders began to accelerate down the runway, no more than a dozen feet apart.

Once airborne, the section raised landing gear, and then retracted flaps and slats in unison before entering the clouds at 4,500 feet. The Intruders continued a gradual climb, flying closely together to maintain sight while accelerating. Passing 9,000 feet, the flight was instructed to switch radio frequencies and contact Seattle Center.

The pair of A-6s broke out of the clouds into a sunny sky at 10,000 feet. All was normal. That abruptly changed passing 12,000 feet, when both Starling and Eagle were startled by a loud "thump" coming from the back of 510. The two men immediately scanned 510's instruments. A second later, the "Backup Hydraulic" light illuminated on the master caution panel. This indicated that the small, backup hydraulic pump was operating due to the failure of one of the Intruder's two main hydraulic systems. Chris Eagle looked across the cockpit to the small hydraulic pressure gauges just above his pilot's left knee. Each of the main hydraulic systems had two pumps, and both flight system hydraulic pumps were pegged at zero pressure.

The A-6 only needed one of its two main hydraulic systems to fly: either the flight or the combined hydraulic system alone could provide sufficient control authority for just about any regime of flight. Still, the moment the flight hydraulic system failed, the training portion of the mission was over. Standard operating procedure dictated that the aircraft with the mechanical problem immediately return to the airfield for a precautionary landing. It was also standard procedure for a wingman, if available, to accompany the damaged aircraft to provide an outside set of eyes if needed.

At the first indication of trouble, Denby Starling keyed the radio and spoke directly to Ray Roberts in 502, using his squadron call sign: "Raybo, we've got a hydraulic failure—I've got the lead."

Taking over as the flight lead would lessen Starling's piloting workload, allowing him to troubleshoot 510's emergency without fear of a midair collision. Ray Roberts eased back on the throttles, sliding aft into a wingman's position.

As 510 assumed the lead, Rivers Cleveland reached to 502's center console and rotated the transponder switch from "on" to "standby," effectively making 502 go invisible to the approach control radar systems. At the same time, Chris Eagle turned 510's transponder knob from "standby" to "on." Air traffic control was now following the highlighted radar return of 510.

Chris Eagle then radioed Seattle Center to let them know that the flight had encountered a problem and required direct routing back to N.A.S. Whidbey Island. The two Intruders turned back toward home base and started a descent. Denby Starling reached down with his left hand and began dumping fuel to get 510 to an acceptable landing weight.

Starling stopped 510's descent 6,000 feet above the water as Eagle requested to hold overhead Smith Island, the designated fuel dumping area. This altitude would allow the fuel to evaporate before reaching the water. Five-ten began a wide, left-hand orbit over Smith Island.

The Lost Intruder

Meanwhile, Ray Roberts was struggling to fly slow enough to provide a visual check of 510. This was not easy to do. Because 502 did not urgently need to land, Roberts opted to save fuel and not turn on his dumps. This put 502 at a significantly higher gross weight than 510, requiring Roberts to fly at a faster airspeed than his lead to keep from stalling. Ray Roberts flew gentle "S" turns slightly behind and to the right of the slower 510 to compensate. Once at the appropriate landing weight, Denby Starling secured 510's fuel dump switch and began to further reduce airspeed. Chris Eagle reviewed the "Flight Hydraulic Failure" procedure outlined in his emergency pocket checklist. As the two Intruders continued their first circle over Smith Island, Ray Roberts reported that a steady stream of reddish hydraulic fluid was running down 510's aft fuselage.

When one of an A-6's two main hydraulic systems failed, the jet's remaining hydraulic fluid was instantly sequestered for use by the flight controls only. This design isolated any leaks in the bad hydraulic system to ensure the ability to keep flying. The physical separation of hydraulic fluid meant that all non-essential for flight—but necessary for landing—aircraft systems needed to be operated via their alternate or "backup" method. The backup system to lower the flaps and slats used an electric motor. Flaps and slats are high lift devices that allow an aircraft to fly relatively slowly for landing. Five-ten continued to decelerate. Approaching 180 knots, Starling moved the backup electrical flaps/slats toggle switch to the "down" position. With the high lift devices now extended, 510 slowed to approach speed as her crew prepared to "blow down" the landing gear.

The two, main landing gear struts located at the base of each A-6 wing were normally lowered by gravity. Each main landing gear and corresponding landing gear door, however, had a mechanical up-latch that required hydraulic power to open. Once unlatched with gear doors open, the landing gear would drop with gravity, locking down into place. Lacking hydraulic pressure, the only way to lower the landing gear was by blowing open the latches using compressed nitrogen stored in high-pressure cylinders. Denby Starling reached for the landing gear handle, rotated it clockwise, and pulled it straight out three inches to activate the landing gear blow down system. Knowing that they had just one shot at blowing down the landing gear, Starling kept the handle extended for well beyond the prescribed minimum of five seconds.

Nothing happened. There was not the expected vibration of the landing gear doors opening in the wind stream; not the clunk of the main struts falling with gravity and the over-center locks engaging. Nothing but the eerie rush of air passing over the fuselage. At that moment, 510's predicament escalated from a relatively routine mechanical malfunction to a life-threatening emergency.

With the main landing gear stuck in the up position, there was no choice but to crash-land the A-6 on its belly. This was an extremely dangerous maneuver that risked cartwheeling the jet and igniting thousands of pounds of fuel in a rolling fireball. The only thing preventing this from happening was Denby Starling's skill as a pilot. Starling and Eagle calmly reviewed the "flight hydraulic system failure" checklist again, step-by-step, to ensure that they hadn't missed anything. But both aviators knew that with the compressed nitrogen blow down bottles now empty, their one shot at anything approximating a normal landing was gone.

Approaching the end of the first orbit over Smith Island, Denby Starling momentarily leveled the wings, pointing 510 east toward the airfield. Commander Starling might have looked up at the runway five nautical miles off the nose, perhaps experiencing a fleeting frustration at having come so close to a near-normal landing. Professional aviators learn to accept reality quickly, however, and Starling probably felt no such thing. The reason why the landing gear blow down system didn't work no longer mattered. They resumed a left-hand orbit over Smith Island.

Chris Eagle read the "main landing gear up" checklist out loud to Starling and then did it again a second time. Rivers Cleveland confirmed over the radio that 510's main landing gear was indeed still retracted, with both gear doors closed flush with the fuselage. The only good news was that the nose gear had partially lowered and was about halfway down. There was a hand pump inside the cockpit designed to get this most critical landing gear component to fully extend. Chris Eagle reached for its handle along the center console and began to work it back and forth.

Five-ten turned past a heading of north, still flying at a landing speed of 120 knots, or 144 miles per hour.

Starling and Eagle quickly discussed what they would do on the ground if confronted with the high probability of a fire, possibly while still strapped into their ejection seats. None of the options were good. The two could eject after touchdown, but that ran the risk of being sucked back down by their parachutes into the fireball. The alternative was to explosively jettison the canopy—their only protection from the flames—unstrap, and try to fight their way through the inferno.

Just about the time that 510 would have come to a stop on the runway if the landing gear blow down had worked, things got worse. Five-ten's two remaining main hydraulic pumps began to dip erratically, while a grinding sound filled the cockpit as the dry pumps began to come apart.

The stick started to feel mushy in Starling's hand. He turned to Chris Eagle and said:

"Get ready to get out of here."

A moment later, Commander Starling pushed the control stick all the way to the left. The Intruder, in defiance, began a slow roll to the right. Realizing that he was no longer in control of the aircraft, Denby Starling gave a terse final command to Chris Eagle:

"Go ahead and get out now."

Eagle grabbed the lower ejection handle between his knees. He leaned back in the ejection seat with elbows in, chin tilted up, and heels on the deck. With his right hand on the ejection loop handle and his left hand squeezing his right wrist, he pulled until the handle moved an inch upward. The ejection was initiated. Chris Eagle's world instantly went black.

I turned to the last page of Chris Eagle's statement to read his description of the ejection.

"I looked down and saw a bubbling white circle in the water which I assumed to have been where the aircraft impacted. It looked to be ¼ to ½ miles south of Lopez Island. I looked slightly to my left and saw that (Starling) had a good chute but was surprised to see that he was lower than I was in the air. I thought of trying to land on Smith Island for a moment, but it was too far. I looked down and saw the water approaching, so I returned my hands to my (parachute quick release) fittings and forced myself to wait until my feet hit the water until I released them."

Naval Aviators are trained to release their parachute the moment they feel their feet hit the water. This is to avoid misjudging the altitude, possibly leading to a deadly fall. Waiting too long to release the parachute, on the other hand, ran the risk of in-water entanglement in a mass of shroud lines, probably leading to drowning. Eagle's statement continued.

"I boarded the raft, got out my survival radio and tried a call on guard. I heard nothing. I made another call, 502 answered back at which time I told them I was okay. Next, I heard Commander Starling call that he was okay and in his raft and ask about me. Commander Starling then asked where the helicopter was and 502 replied that it was launching at that moment. I sat back and tried to orient myself and realized that I was facing the base and about a minute later saw the helicopter on its way out."

"As (the helicopter) approached, I could see a crewman leaning out a port side window, I waved to them, and they went straight over to (Starling). Watching the helo it appeared that (Starling) was less than 300 yards away from me. They retrieved him and came over to pick me up. A swimmer was dropped into the water, I told him I was okay, and he came over and hooked onto my D-ring. I then rolled into the water with him waiting for the helo to lower its hook. After it came down, the swimmer hooked us up, and we rode up together. Once in the helo, I wrapped up in a blanket for the ride to the hospital."

When the search and rescue helicopter touched down at the Naval Air Station's hospital helipad, both men were in mild shock, shivering uncontrollably from hypothermia, but otherwise no worse for wear. The dark, cold waters had released the two lives without argument, apparently accepting the tribute of the Intruder in their stead. I searched my memory for an image of one of the fortunate aviators in a human moment, out of their stoic, military bearing. My imagination settled on a fuzzy remembered vision of Denby's wife and three children, all impossibly young, tearfully thankful that the aviators made it home.

I stood up and stretched. It felt as if I was overlooking something, some hidden message or clue. Temporarily stumped, I turned away from the desk.

Eight

...A PATHETIC ATTEMPT AT A ROUT

May 2014

The extra daylight hours of the Pacific Northwest spring were a salvation. I dreaded the night. Without the sun to warm my core and spark activity, the hours of darkness dragged by with terrible slowness. Alternating between a wrack of frozen pain and dyskinesia's horrible writhing, continually incised by an icy chill in the pit of my stomach, each night threatened despair. The cumulative effect of too little sleep and around the clock fidgeting burned so many calories that, at first, I could not eat enough food to arrest my weight's downward plunge. By late spring, I had dropped to 170 pounds, 30 pounds below my average size since college, and getting lighter. While my muscles had become shockingly defined, a gaunt face betrayed the unhealthy nature of the weight loss. The disease began devouring muscle to sate its appetite as Parkinson's attempted a knock-out blow.

Sundown brought the seductive musings of sinister demons. As soon as it was dark, not until 10:00 pm in the Pacific Northwest, I would collapse into bed exhausted. With the pain temporarily masked by fatigue, I could sleep, but for no longer than an hour. Then came the kicking legs. Restless butterflies fluttered in the pit of my stomach, scratching at the back of my spine with leathery wings until the pain of stiffening joints jolted me awake.

For the rest of the night, no dreams, no comfort, no real rest. It was night after night of endless hours back and forth, from the bedroom to the kitchen and back, trying to sleep. The only way to ease the sorely stabbing pain was by walking; pacing laps around the empty living room before heading to the refrigerator. The night's first "snack" might be a heaping plate of leftovers. Finished eating, I would brush my teeth again and lie down. The process would repeat four or five times until, surrendering in frustration, I would stay awake. Eventually, my body reclaimed five hard-earned pounds through disciplined persistence. The night came to mean an endless prowling angst, fighting to literally not wither away in a time-distorting exhaustion.

Finally, saved by the 5:00 am dawn, the glowing horizon would melt the evening's lonely anguish; in a flash, the dark terrors were temporarily forgotten. The anxious night shadows were a poor substitute for REM sleep, but feeling comparatively refreshed nonetheless, each morning brought temporary freedom with the knowledge that I could drive at least a short distance. After mid-morning, my ability to safely operate a vehicle was hit or miss, and never for far. Borrowing from a pilot's phrase relating to cabin depressurization and hypoxia, a condition of insufficient oxygen in the body that can lead to lightheadedness, passing out, and even death, I'd come to think of these precious morning hours as my "time of useful consciousness."

The darkness used to be a friend: the Intruder's mission required long, dangerous flights over enemy territory, "feet dry," with night's cover a welcomed ally. Ironically, the darkness, now clearly on the disease's side, connected me to the missing A-6. The coming day's new trials inspired me to survive the lonely night. When not actively searching for 510, a surrogate activity would be found or invented to bridge the gap. The strategy was simple: aggressively accumulate small victories in the day to bolster morale during each long darkness.

I was proud of these battles where the measure of success was in not surrendering, and perhaps that reliance on pride should have been a warning. I had experienced this confluence of ultra-awareness before, but only for brief moments, with a whole mind and body. At the pointy end of the spear in 1991, during the opening night of Desert Storm, in the pilot's seat of an A-6 Intruder. And in 1983, lost deep within the wreck of the *Andrea Doria*'s First Class Dining Room, looking for the exit and survival.

The cause of Parkinson's disease is unknown. There is some speculation, however, that it might be brought on by exposure to high oxygen partial pressures—such as from deep diving—or neurological toxins. I had flown through thick, ugly clouds of oil fire smoke and possibly residue from bombed chemical munitions during the first Gulf War. It struck me that the two defining experiences of my life, diving and flying, might not be finished with me quite yet.

A refusal to quit became a hallmark of daily ritual. It probably appeared to friends as senseless stubbornness. They would watch me perform an oil change in *Sea Hunt's* tight engine compartment, a ninety-minute task of old, now complete only after five hours of twisting, non-stop perseverance. More than once I found myself between the engines after sunset with flashlights staged, struggling to install a final component, at times taking two hours to secure a single screw. It must have been hell to watch. But from my perspective, the process transformed my loss of capability into an exercise in concentration. It was an intense, bizarre sort of fun.

We had found the promising contact on the first day of the project's side-scan sonar search for a reason. I was meant to dive the A-6 of my old squadron and memories, and there was no question in my mind that this first contact was 510. The sonar contact's depth, 145 feet, was a quantifiable target in a world of intangibles. It was not in the same league as my pre-Parkinson's diving depths, but that made no difference. This was a new competition in which old rules no longer applied. These were simply two different phases of my life, and that they both involved diving was incidental. Yesterday's challenge was overcoming an obstacle with a particular outcome, a goal, in mind. Today's task possessed intrinsic value, and nobody else's opinion mattered. I was the only judge.

In some ways, today's underwater trials were infinitely more dangerous than the dives of old. How would my body react to a depth of 145 feet? Would the disease, my medication, or an obscure interaction between the two cause me to black out? Would my dexterity improve with depth or suddenly deteriorate without warning? Disturbing visuals peeked at the edge of my consciousness: losing the regulator from the grip of my teeth, staring at unresponsive fingers as I began to drown, unable to return the life-sustaining gas source to my mouth. I saw my still body lying heavy on the black, abysmal bottom, unable to manipulate either the weight belt's buckle or the buoyancy compensator's inflator mechanism. The images only haunted me when far from the water, however. Once approaching the dive site, confidence from purposeful activity returned. Such catastrophic human failures are exceedingly rare in diving and are almost always accompanied by prominent warning signs. It would probably be the same irrespective of the disease and the medications; probably.

Still, before experimenting at 145-feet deep, 50 feet deeper than attempted since diagnosis, I had to be damn sure that my proficiency in basic diving skills was second nature. My thoughts returned to the previous spring, the last time I had felt completely at ease underwater. It had been a vacation to Port Hardy, British Columbia. The Canadian experience taught me that if I could just work through the discomfort of several warm-ups, then diving could become almost as comfortable as enjoyed decades earlier.

The eight-day vacation to the northern tip of Vancouver Island offered three boat dives a day. After pushing myself to a second daily dive soon after arriving, my energy level remained surprisingly high. I even tried a third dive as the end of the vacation approached. There was a point, however, where failure was sudden and unforgiving. Three dives a day was not safe even back then, a full year earlier.

The moment after receiving Ben's email describing the promising target, I literally threw my dive gear in the back of the Expedition, drove the ten minutes to the marina, and got suited up for the unceremoniously slimy work of cleaning *Sea Hunt's* hull. I took satisfaction in managing to gear-up without assistance. Working up a sweat after ninety minutes of non-stop scrubbing, I finally climbed exhausted up *Sea Hunt's* dive ladder unaided.

Cleaning a boat bottom is simple diving but hard, dirty work. After hosing down my gear with fresh water on the dock, I felt bone-tired but oddly energized, a most welcome contradiction. I had stopped personally cleaning *Sea Hunt's* hull several years earlier, instead choosing to hire out the chore. It was galvanizing to know that my biggest obstacle might be a psychological barrier; I was growing accustomed to the pain. The experience motivated me to complete the monthly back-breaking drudgery for the remainder of the spring and summer. Nobody likes an uncontested beating: I fought back.

Hot yoga became a passion, and between increased core strength, flexibility, and countless pushups, my confidence grew. To move through poses so mindfully, focused and aware, helped me to accept all the extraneous motion over which I had no control. And the meditative aspect of yoga offered relief to a brain under constant assault, temporarily providing comfort from what was becoming increasingly obvious to me was an insane world. It helped me find a semblance of peace when my body refused my mind's commands.

Exercise also allowed for a detouring of frustration to a single-minded determination. Whenever the dyskinesia became too great and threatened to throw me off task, I would rise stiffly from my chair and force out a painful set of pushups to help concentrate my body. The second set came easier, and by the third, with music blaring and affirmations of "I can do this" muttered under my breath, my determination would be renewed. The constant battle put me in, what felt like during brief moments of fantasy, the best shape of my life.

From a physical perspective, it was entirely self-delusional: all my efforts were at best marginally slowing the disease's acceleration and the wasting of muscle. Still, each workout bolstered my spirit, earning me one real, hard-earned trait: I was undoubtedly the toughest I'd ever been.

Delusion aside, it was clear that time was not my ally. I needed to accomplish the practice dives quickly. I had not made a decompression dive since diagnosis, almost a decade earlier. These were serious concerns. Would the disease make me more susceptible to the bends? And what of the possibility of the disastrous convulsions of oxygen toxicity? I carefully shuttered away such thoughts in a dark corner of my mind.

I was in the water for a second warm-up dive the following day with a friend I had met on the Port Hardy vacation. We spent 45 minutes in the shallows of the Keystone Jetty, mid-way down Whidbey Island. I struggled for part of the time underwater, but only when compared to the comfort level of the Port Hardy vacation.

The next day, Thursday, May 15, Ben and I were back onboard *Sea Hunt* to refine the contact's position and finish surveying Lopez Island's southern coast. This would be our only search day without a dedicated deck hand. Ben moved continuously between the cabin laptop and the hand-spool of transducer cable on the stern for six hours of intense, continuous effort in some of the area's worse currents. The cumulative fatigue from the practice dives earlier in the week coupled with the challenging currents pushed me to the edge. The experience should have taught me to pay more attention to my rapidly changing limits, but pride is a myopic regulator of confidence.

My erratic performance at the boat's helm must have been confounding for Ben to watch. I had not offered a detailed explanation, thinking it presumptuous of me to do so. Damn good at hiding my discomfort, I would soon learn that even some of my best friends had little understanding of what a struggle it was for me to make it through the day. The physiological changes to my body were simply beyond the ken of most people. And why wouldn't they be? What life experience was necessary to instill an individual with the proper balance of empathy, intuition, and knowledge to allow them to really take a good look around, to stand in my shoes? I could not come up with an answer, so how could I possibly expect anyone else to be genuinely thoughtful about my circumstance?

The day's demanding sonar runs heralded an impending turning point. The towfish was not heavy enough to descend beyond 200-feet deep while being towed at the proper speed. The only practical way around this problem was to somehow outfit *Sea Hunt* with a winch powerful enough to haul in a new towfish with a heavier cable, one that could accommodate additional clip-on weights as well. Secretly, I considered the deep-water sonar and winch unnecessary. But Ben's existing plans included purchasing a more advanced sonar system at some point, and the unit he had in mind was sold only in Sweden. It would take time to ship, probably delaying the project a month, more than enough time to identify the 145-foot contact as the missing A-6.

The Lost Intruder

While Ben investigated buying the deep-water sonar, I made three additional warm-up dives, all with double tanks, the type of rig I could expect to use on the contact identification dive. By gradually increasing the depth of each dive, I was soon able to descend to 135 feet for several minutes before starting up for a practice decompression. The training honed my fundamental skills, but the unknown physiological effects of the greater pressure were still troubling. And just a single dive left me exhausted for the remainder of the day, and occasionally the one to follow.

The prospect of an increased susceptibility to oxygen toxicity was especially disturbing. Convulsions could come without warning, potentially at the deepest, most dangerous point of the dive. Was it worth the risk? I struggled with the answer. In the end, it wasn't just the dive that was at stake. Was my commitment to the project sincere, or was I fooling myself? With no clear answer, attempts to dispassionately judge the hazards had the counterproductive effect of increasing the pressure put on me, by me, which further aggravated my symptoms.

To back out of the dive, in my mind, would be to cave to Parkinson's most relatable challenge. The importance of finishing an oil change did not translate to most people: why not just hire out the chore or ask for help? But backing out of a dive was a different story. To accept defeat in an early test so fundamental to the project would be devastating, an irrecoverable psychological stumble in the longer-term fight, which, in my mind, meant death anyway. But where did sensible resistance end, giving way to the motivations of an unbalanced ego? I didn't know.

Most of my warm up dives were done solo. This decision mirrored my approach in fighting the disease, of placing the highest premium on self-reliance. I had tried support groups but found them to be counterproductive bitch-sessions. I would leave these meetings more depressed than when I had arrived. My battle with the disease was solo by necessity; the transfer of this attitude to diving seemed to be a natural fit.

The MDS crew were also anxious to get "eyes on the contact," and we scheduled our first dive attempt just one week after discovering the promising target. Washington weather can be unpredictable, however, and the wind picked up suddenly during the morning of the planned descent. Not only would the white caps make it difficult to find the contact with the Dragonfly sonar, but the waves could cause the transom to come crashing down unexpectedly, potentially on a diver's head while exiting the water. I called the MDS crew and called it off. Scrubbing the attempt was a positive decision in one critical respect; it allowed the time for more practice dives. The next attempt to identify the contact was set for two weeks later. In frustratingly similar circumstances, the winds again started to build in the morning. The MDS divers all lived hours away. Not wanting to waste the day for them, I canceled early.

At a little past noon, the winds abruptly subsided and the seas calmed. I scrambled to put the dive back on for the next slack tide, six hours after the original plan, but the MDS divers had scattered for family commitments. By early afternoon, conditions were perfect. Growing frustrated, I called a friend, Dan Crookes, to see if he would be willing to operate the boat during the late afternoon slack tide while I dived solo. He said yes without hesitation.

For 35 years, most of my diving has been done alone. While I had kept up with the technological advances in diving over the years, the disease drove me to be selective in choosing between old and new technique. There were good reasons for the older style dive practices that I did retain. The increased rigidity of my shoulders made it impossible to reach the manifold that connected the two tanks behind my head. There was a total of three valves on the common manifold connecting the two gas cylinders that needed to be accessed in the event of an emergency. Although hot yoga had increased my flexibility, it was a temporary limberness that quickly vanished once I stopped stretching.

Looking back to experience, I placed a separate regulator on each of the main tank valves, planning to switch back and forth periodically to balance the remaining gas. This was accepted practice when I'd first learned to wreck dive. It basically guaranteed at least a third of a full tank in an emergency. The advantage of the new style of tank manifolds was the ability to continue to draw gas from both tanks if one regulator failed. But that required the isolation of each cylinder in the event of an uncontrollable free-flow, a procedure I was now unable to accomplish because I could not

The Lost Intruder

reach the manifold. I'd safely dived with this "old" configuration of independent tanks to far greater depths, hundreds of times. Of course, back then I had been healthy.

By the time *Sea Hunt* reached the latitude and longitude coordinates of the sonar contact, the wind had died down to nothing. We made several slow runs across the coordinates until we had identified the sonar image of a huge boulder. With *Sea Hunt* hovering over the northern tip of the rock, I dropped an anchor attached by a 160-foot line to a large, white buoy-ball. The plan was for me to descend along this downline while Dan Crookes maintained *Sea Hunt* in position overhead. Once finished on the bottom, I would ascend to the surface along the same rope.

I took my time suiting up, double checking the proper function of the two independent gas sources, three lights, three knives, and two computers. Satisfied with my medications for the next hour, I continued to get dressed. After pulling on a neoprene hood and donning my mask, I added one relatively new component to my diving routine by strapping a GoPro video camera to my forehead. Video documentation would be essential if the contact were indeed the missing A-6.

Dan Crookes moved confidently between *Sea Hunt's* aft deck and helm. Dan, the owner-operator of a 100-foot landing craft, was a well-seasoned mariner. Propane and gasoline were not permitted on Washington State ferries, and his vessel, the *San Juan Enterprise,* transported these combustibles from the mainland to customers on the islands. He also ferried construction materials and general supplies to those islands without regular transport service. Having Dan watch the boat while I was underwater eliminated a lot of worries. If something unforeseen was to happen on the surface, I was confident Dan could handle the situation.

I turned on the GoPro's recording function and prepared to jump in. The rock being used as a reference point was much bigger than an A-6. The contact was supposed to be at its base, but I didn't know which side. Unable to get a hold of Ben and concerned about missing the slack tide, I decided to search the northern, slightly shallower rock perimeter. I told Dan that I was ready. A minute later, the white downline buoy appeared twenty feet off the stern.

Clamping down on the regulator mouthpiece with my teeth, I placed my left hand firmly over both mask and the GoPro before making a giant stride entry off the stern. Once clear of *Sea Hunt*, I turned with a flip of my fins to face Dan, then nodded in the affirmative, a signal of "I'm okay." I kicked on my back to the buoy-ball.

Approaching the downline, I stopped kicking and slowly allowed my feet to settle until I was vertical in the water. Going still, I took several deep breaths to lower my heart rate from the surface swim before depressing the buoyancy deflator button. Bubbles streamed from the corrugated hose leading to the bladder on my back. *Sea Hunt* quickly faded from view as my fins dropped into the darkness. Exchanging the bright sky for the soupy green of the shallows could be unnerving for a second, and I focused on the reference of the white downline. Passing ten feet deep, two quick pumps of gas back into the buoyancy compensator slowed my drop as the water pressure squeezed the bladder smaller. With my descent now stabilized, I continued down at a controlled pace, allowing the belted lead weights around my waist to drag me to the bottom.

The downline trailed into the depths, disappearing into nothingness. I glanced at the wrist computer on my right hand as the digital readout passed 50 feet. It began to get dark. Passing 70 feet, I reached for the handle of my primary dive light. The water's greenish hue succumbed entirely to blackness at 90 feet.

I paused, allowing ten seconds for my eyes to adapt before resuming the descent. I pointed the light directly into the murky oblivion while straining for the first glimpse of the boulder. My imagination exaggerated the small possibility that the rock was shrouded in a fishermen's lost net, waiting for me to descend into its deadly, drooping folds.

The downline lessened its angle and looped as the last vestiges of current disappeared. *Sea Hunt's* rumbling diesels idled loudly overhead. At 120 feet, the dive light pierced the darkness, exposing a two-foot spot of marine growth on an encrusted, jagged boulder of massive proportions. I scanned the light along the rock and dismissed concern for a lost net. Visibility was awful. The dive light barely penetrated four feet in the surrounding blackness before its beam was completely reflected by the suspended silt. Each small landmark immediately disappeared once out of the light's direct beam.

Concentrating on the measured resistance of each slow, deep inhalation, I cherished the reassuring tension of the downline between my fingers. I hovered neutrally buoyant at 125-feet deep, my right arm cradling the thick rope while eyes strained to see the bottom. I was perched on the edge of a plateau leading north to a sharp drop off. Nothing but solid rock all around except for north, where the downline disappeared after four or five feet. How much further to the bottom, I thought? Could there possibly be enough room between the plateau and the sea floor for an aircraft? I felt for the second regulator that ran to the full left tank on my back, switched it out with the one in my mouth, and continued to breathe.

The warm-up dives had shown me that, while it was difficult getting suited up due to a lack of dexterity and finger pain, once underwater things improved significantly. Perhaps because of the distraction to my focus, the pressure even seemed to alleviate some symptoms. Floating neutrally buoyant, the joint pain was almost entirely gone with the apparent absence of gravity. The only persistent source of discomfort was an involuntary tendency to scrunch up my face, causing my mask to lose its seal and leak. Clearing a flooded mask is easy, but constantly having to do so can be a dangerous nuisance.

I cautiously followed the line to the bottom. The heavy chain leading to the anchor rested in the mud absent a current, allowing for the release of about twenty feet of slack rope. I took advantage of the extra line, grabbed the rope in my left hand and swam it out 90 degrees to the direction of the anchor.

Seeing nothing on either side of where the rope turned to chain, I continued toward the anchor. At 135-feet deep, it was a flat, sandy bottom with occasional mud patches. No fish swam into the dim light. There were none of the white Plumose Anemones of well-circulated water, nothing but nondescript marine fauna—barnacles, greenish seaweed—clinging to the muddy surface of the enormous boulder. And there was definitely no A-6 Intruder.

Fifteen minutes went quickly, signaling that the bottom portion of the dive was over. On edge for the entire evolution, I never relinquished a loose grip on the downline. I started a slow ascent, spending an extra ten minutes at fifteen feet to help compensate for any increased decompression vulnerability. Finally satisfied that it was safe, I inflated the buoyancy compensator and broke the surface with left arm extended.

Dan immediately caught sight of me from *Sea Hunt's* deck fifty feet away. Ten minutes later, with Dan's help, I was out of my dive gear, pills in hand, ready for my next dose. From freedom to stranglehold, life moves just that quickly.

It had been a dark and lonely dive. That was the intangible aspect of solo diving—the spooky factor. While the gloomy isolation had never kept me from diving before, I grew concerned. Eerie conditions ratcheted up the stress level of any dive, and if there was one thing I did not need, it was more anxiety. It felt as if I had barely escaped an unseen snare, never realizing just how close to the end I had come. The next dive, I resolved, would be made with buddies.

The author waves a gloved hand at Dan Crookes on
Sea Hunt after the solo identification dive
(from the author's collection).

Nine

Eyes locked in mortal-gripped vision

May/June 2014

The solo dive imparted a better appreciation for the confusing underwater terrain, but did little to help me identify the few distinct boulder features captured by Ben's side-scan sonar. From the south of the target, the side-scan image revealed nothing to the untrained eye except the scattered electronic static of bottom clutter. The other two perspectives were clearer, but it was only after the specialized software "stitched" the three side-scan angles together that the stick figure of an A-6 magically appeared.

I spent Friday on preventive maintenance and readying the boat for the next day's dive. Dan Warter and Rob Wilson, the only MDS divers available on Saturday, would both be using rebreathers, but they seemed comfortable having me tag along in my older style, open-circuit scuba equipment. I'd first met Rob in 2000 after a Mukilteo training dive with Ron Akeson. Rob was out for a ride on his motorcycle when he saw Ron and me with our gear sprawled across two picnic tables at the edge of the parking lot after the dive. Rob had been midway through the trimix course at the time but was taking a break from the syllabus for work commitments. He seemed to be an easygoing, friendly guy back then, and he lived up to this first impression in all future encounters.

Rob's maroon pickup pulled into the Deception Pass Marina parking lot at 7:00 am with Dan Warter in the passenger seat. Ben followed fifteen minutes later. Rob and Dan immediately began humping their gear down the ramp, donning their heavy rebreathers for the walk to where I had moved *Sea Hunt* adjacent the fuel dock. Rebreathers had come a long way since I had last seen one more than a decade earlier.

Each rebreather was connected to two small bottles mounted on an integrated backpack, one filled with trimix, the other with pure oxygen. Rebreathers worked by feeding the diver a metered proportion of pure oxygen with a diluent blend of trimix. As the diver exhaled, the used breath would cycle through a carbon dioxide scrubber in a closed circuit before joining a recalibrated ratio of oxygen and trimix. The new, reformulated mixture was then ready for the next inhale. The divers would also carry several full-size stage bottles, clipped one or two to the diver's side, to serve as emergency bailout tanks if a rebreather failed. For practical purposes, the rebreathers extended a diver's time underwater almost indefinitely.

With my equipment already aboard, I helped load the rest of the gear onto *Sea Hunt*. Ten minutes later, we shoved off and I turned the helm over to Dan Crookes so I could listen to Rob brief us on the conduct of our underwater search. The plan was to drop the downline anchor at the southwest edge of the boulder, on the opposite side of the rock from my solo dive. Once on the bottom, Rob would reference a compass heading to follow the perimeter of the rock to the north. The A-6, if it was there, should not be more than about fifty feet from the anchor. Dan would follow with the bright lights of his video camera. Once we had explored the boulder's northwest quadrant, I would break off from the others and ascend, barely incurring a decompression penalty. If nothing was found, Rob and Dan would continue the search along the periphery of the rock's southwest corner.

I had bought a new light, one bright enough to triple my field of view, that also conveniently strapped to the back of my hand. Despite being a slightly deeper dive, the new light and company of dive buddies made it a far more relaxed experience than my solo venture. All we saw to the north was the craggy outcroppings of the same, massive rock. The sonar contact, whatever it was, must lie to the south, I concluded, as we turned around.

Rob had snapped a thin, nylon guideline to the anchor chain at the beginning of the dive. I swam ahead of Rob as he reeled in, letting the white return line slip through my fingers until reaching the anchor. When we reached the downline, I raised my hand in a wave of goodbye and started my ascent. The lights of the other two divers were still visible below me as they headed to the south for the remainder of their search. I wished them a silent good luck.

I took my time ascending, made a conservative decompression, and broke the surface fifteen minutes later. Ben was looking out over the water's surface from the aft deck while Dan Crookes worked the controls at the helm. After swimming the short distance to *Sea Hunt*, I placed my right fin on the dive ladder's bottom rung before raising up to reach for a handhold. With help from Ben and Dan Crookes, I stood up on the swim platform. Ben kneeled to remove my left fin as I made my way through the transom's cutout. Twenty seconds later, I was sitting on the cooler with my tanks lying down beside me.

"How do you feel?" Ben asked, looking me up and down slowly, stopping at my eyes, scanning for signs of a dive-induced neurological problem.

"Great. No A-6, but I'm okay." I started to make a wise-crack but reconsidered. Ben was at one time a Seattle firefighter and paramedic. If he cared enough to ask, then I figured he was owed a serious reply.

I scanned the water looking for bubbles before remembering that Rob and Dan Warter were using rebreathers: there would be no visible exhalations, no way to track their progress across the bottom. Twenty minutes later, two hooded forms broke the surface. The pair of divers began a measured swim back to *Sea Hunt*. Dan Warter still had his rebreather on his back when he reported from the dive platform that the sonar contact had been an oddly shaped rock pile. A nearby, water-saturated log had completed the three-dimensional sonar image of an airplane. We were back to square one.

It was patently obvious to me that there was something fundamentally wrong with my theory's search area. A fresh outside perspective was needed, preferably from someone with a technical knowledge of the A-6. It was still possible that my postulated ejection point and final course were correct, but we were running out of unexplored water that would allow for even the most generous interpretation of my hypothesis. And most of the unsearched depths were more than 200-feet to the bottom.

It was fast becoming unlikely that 510 would be found in water shallow enough for me to take part in the next identification attempt. While my dive with Rob and Dan had been reasonably comfortable, the residual effects ambushed me later that afternoon. With my medication timing and dosage out of whack, I was assaulted by an almost violent bout of dyskinesia. The next morning, after a sleepless night of muscular rigidity and joint pain, my entire body ached from toes to ears. I would entertain, maybe, a dive to a repeat depth, but absolutely no deeper.

The dosages and timing of my five prescribed Parkinson's medications had come to need constant adjustment. My neurologist, a movement disorder specialist, was great, but she—like everyone else not suffering the disease—didn't have a clue on a personal level how Parkinson's felt. For the first five years after diagnosis my symptoms were relatively minor, and the biannual neurologist appointments were all the same. Each doctor visit used a series of questions to decipher what was working, what wasn't, and how best to tweak the drugs to maximize effectiveness. This information navigated the filters of my best description of a symptom, and then the doctor's attempts to understand how best to mitigate the disease's effects. Recommended changes to my drug regimen would follow, a trial and error process by necessity.

As the disease progressed, I realized that a more responsive strategy would be to "cut out the middleman" and take control of my own treatment schedule. I still relied entirely on the doctor's advice regarding the safe limits of each medication, as well as other treatment options and anything new on the horizon that might hold promise for me. But all pill dose and timing decisions, within the doctor's broad parameters, were completely in my control. If a drug regimen started to lose effectiveness, I didn't have to wait weeks or months for a neurologist appointment. Instead, I experimented incrementally until finding a better combination. This approach would seem to benefit all Parkinson's patients, but in my limited experience very few are willing to accept responsibility for their own treatment.

Altering my drug regimen was not the only treatment variable that I experimented with. Daily exercise was a must if I was to maintain the energy for any semblance of a normal life. This might mean simple calisthenics while out on the boat or yoga and the gym when ashore. Vigorous exercise also kept Parkinson's muscular rigidity and depression from reaching debilitating levels. It wasn't easy getting started on a workout, and worsening joint pain required frequently looking for new types of exercise or a workaround, such as doing pushups on my fists. But it was always worth it in the end.

Maintaining a positive attitude was another critical variable. I found mantras to be a helpful tool for staying optimistic, particularly when concentrating on a delicate task or working through pain. Thoughts follow words, and the effect is real. Seeking new challenges during those brief moments when I could muster the drive also bolstered a positive attitude. Even the simplest repetitive motor activities, such as placing small screws in their holes, provided the opportunity to break through a wall of frustration and clear my mind. It was a meditative good for the soul.

My body's reaction to the interdependence of drug effectiveness, exercise, and mental attitude was constantly in play. Like landing on the carrier, the adjustment of one variable usually required the fine-tuning of the others. The Naval Aviation mantra, "meatball, lineup, angle of attack," helped me to remember these connections. It was only through focused concentration and a surrender to the natural flow of things that an instinctual self-confidence could consistently be achieved. When these variables were in synch, it helped me to sleep better, which strengthened my overall response to the disease as well.

With the search temporarily stalled, Ben and I informally split our efforts. He worked on the sonar and winch issues while I sought out a second opinion on the validity of my theory. I considered the Attack Squadron 145 aviators from the era of the ejection who still lived in the local area. The perfect man for the job came to mind almost immediately: Robert "Tugg" Thomson. For a tour of duty, Tugg had directed the A-6 bureau at the Navy's flight test and evaluation squadron in China Lake, California.

Tugg had a wealth of experience flying the A-6 outside the normal flight regime, offering a rich background that might enable him to see something I couldn't. I called Tugg on June 9th, and we set up a meeting for the next day. Tugg is an avuncular teddy bear of a man who "spoke flying" with his hands like a maestro conducting a rock opera. His presentation of nearly everything was over the top, but his advice was always sound and often creative. As it turned out, Tugg's most helpful suggestion had little to do with flying but was critically important nonetheless. Tugg recommended submitting a Freedom of Information Act Request (FOIA) to the Navy's Judge Advocate General (JAG), the Navy's lawyer corps, for the written investigation of the incident.

There was an excellent reason why I had not yet requested the JAG Investigation: after 25 years, I'd forgotten that it existed. Two separate reports are generated by a significant Navy aircraft accident. The first is the better-known Flight Mishap Report, which bears responsibility for determining accident causation from a safety of flight perspective. The second report is the JAG Investigation, which is tasked with determining whether fault could, and should, be assigned for the incident, and to whom. Because the reason for the total hydraulic failure was never positively established, in the end, responsibility was not assigned to 510's crash.

Navy aircrew are typically required to read an abridged and de-identified version of the Flight Mishap Report after any major accident. This is to learn from the emergency in the hopes of avoiding a repeat incident in the future. The complete, unedited Flight Mishap Report is not open to the public, not even through a FOIA request. Because the report's credibility depends on the absolute candor of all those interviewed, the best way to ensure honest answers is to guarantee that the investigation does not negatively impact any individual careers. All testimony was subject, therefore, to a type of immunity, and the report itself was off limits to the public.

When Tugg asked me if I had considered a FOIA request to obtain the JAG Investigation, I promptly gave him a "what the hell are you talking about?" blank stare. I asked him how the JAG Investigation might help. The answer was simple. The JAG Investigation mirrored the Flight Mishap Report in most critical respects. The two investigations were conducted simultaneously, with each requiring endorsement up the same initial chain of command. If one varied significantly from the other, then the disparity would need to be resolved before it went any further up the command hierarchy. The JAG Investigation should contain the control tower logs, the radar plots of the two A-6s before the crash, as well as the tower radio transcripts, written witness statements, and more. I went straight home to research how to properly submit a FOIA request. The FOIA request letter went in the mail June 12th.

The break in the search schedule while awaiting the deep-water sonar left my focus to wander, a potentially dangerous situation for a man with Parkinson's. To keep my mind actively and positively directed, I took every opportunity to "people watch," carefully observing human interactions, to fill the void. From driving habits to saying hi to strangers on the street, there seemed to be certain commonalities that begged for a broader motivational definition. Why did we do the things we do? My attempts to answer such questions inevitably converged on some event from my past, with one memory standing out from the others.

It was not that long ago when "things" had been important to me: material accumulation proved, in my mind, that I was one of life's chosen winners, which, of course, carried with it the judgment that others around me were losers. What had once been a driving force no longer resonated. It had been a hollow existence, devoid of true compassion or circumspection. But, oddly, I felt no resentment or sadness; no sorrow or regret over wasted priorities and time. This reflection's only negative emotion stemmed from a mild frustration at my inability to convince others that I was now, for perhaps the first time in my life, genuinely happy.

Happiness: that utterly human, elusive state that money really can't buy. The event that triggered this realization was the third annual white water rafting trip with a group of friends in June of 2014. Looking back at the GoPro river video of our first excursion in 2012, it was apparent that the disease had affected me considerably since then. Most people have difficulty looking beyond the physical, however. What was harder to see, but was clear to me, was that the change was for the better.

It reminded me, strangely, of the timeless Socratic paradox: "I know only one thing—that I know nothing." It was, in my opinion, not a sentence about quantity. The meaning did not lie in how much or how little one knows, but rather steered to the neutral acceptance that we humans are wildly ignorant. This is not only okay, but if embraced, it is a concept that can settle one's world-view into a more digestible form. Meaning does not reside on a distant intellectual plane but is in front of us all, most obviously to me in the words of children, those less corrupted by "higher" thinking.

During the three-hour drive home from the rafting trip, Jared, Sean Kelley—a 19-year old friend of the family—and I engaged in one of the most enjoyable conversations of my life. My son was at the wheel, as I was seriously "off" for the duration of the return drive during which I transited three of my Parkinson's states. First came the jagged twist of an hour's worth of a writhing, mumbling dyskinesia of optimism, interrupted by the gloomy descent to dystonia's harshly strained muscles, and just as quickly back to dyskinesia.

One of the difficult to convey positive aspects of the last three years was that I had remarkably little concern for what other people thought about my opinion or me. What was important to me was gaining an understanding of the truth, even when the truth was ultimately unobtainable. Routinely having people stare at you in public with generally negative expressions is a wonderful way to get over skin-deep impressions. It's a good way to get over yourself. The happiness came first; I was just trying to understand its source. I came to four conclusions after that car ride home.

First, the most fundamental human motivation resides in a tribal need to belong, which at its inner core is ego-based on personal insecurities. This is what leads us to compete in the material world of false value; it is what distracts us from honest meaning.

Second, the ability to develop authentic ownership of one's identity devoid of external influence is probably a person's most empowering attribute. To understand one's self, divorced of the husk of false trails and chimeras that the world attempts to foist on us, is critical to a sense of inner peace, without which happiness is not possible.

Third, it is all okay, even—especially—when it is not. Work hard to accept the unacceptably unchangeable. We will never fully understand the world around us or why things happen, so what's the point in worrying about it?

And finally, grow to accept that you know nothing; in fact, revel in it. Be happy.

This epiphany did not come out of the blue; perpetual lack of sleep and constant pain had intensified yoga's meditative element within me. But it was more than that. When I look back to those months pre-DBS surgery, I see an unprecedented awareness—powers of observation previously unexperienced—that I've not as of yet been able to fully recapture. It was the perfect silver lining to a troubled existence. How could events not be planned, not be purposeful beyond our understanding? Didn't we all have a function in some unseen meta-universe?

For the remainder of June and most of July, Ben kept looking for a suitable winch while I scoured the few remaining unsearched shallow spots close to Lopez Island with the Dragonfly. Ben had earlier identified a small contact directly at the eastern end of Colville Island in just 55 feet of water. It was much too small to be the A-6, but there was a long shot that it could be a part of an aircraft debris trail. I decided to dive the site with a friend to identify the contact.

This descent went very differently from the pair of identification dives made just a month prior. It was mid-morning, and my symptoms seemed insignificant. I had slept a few hours the night before and felt good on the boat deck, but once underwater, nothing came easily. It was an uncomfortable evolution, a tortuous fight to keep my mask clear of flooding and my buoyancy under control. Without the dexterity to manipulate my buckles, I needed help getting out of my gear once back on the boat. The dive left me feeling profoundly uneasy.

Nothing obvious had changed from the dives a month earlier: medications, exercise, attitude, my general routine were all the same. Everything was just inexplicably harder. We did find the contact, another out-of-place rock next to an abandoned mooring buoy, but the experience haunted me. Was the disease really accelerating this quickly? The answer floated uncomfortably in my mind on a parallel path to my newfound contentment. Time was running out.

Ten

THE BLACK-SOUL TEMPEST APPROACHES

July 2014

The deep-water sonar order processed faster than expected, reaching Ben just in time to fill an uncomfortably wide gap in the project's July schedule. Acquiring a winch, on the other hand, continued to present a problem. Just the same, for now, we could expand the search modestly into deeper waters. In a rapid exchange of emails, we agreed upon two days to resume the search, Sunday, July 20 and Tuesday, July 22. Needing a deck hand, I asked 16-year old Jared to go along. We planned to work the flight profile in reverse, moving our search along 510's estimated final course line from the most distant possible impact point toward the best-guess ejection coordinates.

We planned to start our side-scan runs in water 180-feet deep to the south of Colville Island. As the survey area moved to the southeast, the bottom sloped downward. If the search were to continue beyond 200-feet deep, then the project would be forced to go on hold until we had access to a winch. Ben correctly estimated that it would take two full days to complete the 180 to 200-foot sector. The search pattern also had *Sea Hunt* skirting the shipping lanes, which posed no great challenge for the time being, as both days were sunny with unlimited visibility.

The Lost Intruder

We got a late start on Sunday due to the need for a minor electronic repair and didn't leave the dock until early afternoon. *Sea Hunt* had barely nosed out into deep water when Ben dropped the towfish off the stern to calibrate the deep-water sonar. Like a kid with a new toy, Ben stared eagerly at the laptop display: it was one mighty expensive toy at a cost of $16,000. Ben had committed a lot of time and money to the project, and he was not charging me a dime. The fact that we had gotten as far as we had was thanks almost entirely to Ben.

As *Sea Hunt* nosed out into the straits, I set us up for several straight runs to calibrate the sonar, not expecting this simple evolution to turn into a monumental chore. After thirty hair-pulling minutes and Ben's gentle prodding, it became evident, even to me, that somebody else needed to give the helm a try. I was thankful that Jared was with us. Passing the helm to my son made the prospect of giving up easier. Jared steered the boat on a near-perfect course for the rest of the day, allowing me to rest up, give him an occasional break, and then take over for the final two hours after darkness had fallen.

We halted Sunday's search after eight hours. Tuesday was another story, with the survey spanning nearly thirteen hours. As we moved further from shore, it became increasingly difficult to get the towfish to drop to its required depth. The good news, from a helmsmen's perspective, was that there was virtually no chance of dragging the towfish on the bottom. Aside from the usual jockeying of throttles to compensate for the current, it was a lazy, beautiful day with the shipping lanes mostly empty.

Lawson Reef is shaped like a jagged "U," with rocks that nearly break the surface at the points farthest east and west. These shallow outer reaches of the reef are marked by red navigational bell-buoys, each sitting two and a half miles apart. Rosario Strait's shipping lanes are positioned to take advantage of the unobstructed water to the west of Lawson Reef and the deep water at the bottom of the "U."

Rosario Strait's two shipping routes operate like highways, requiring ships to stay to the right. There is a one-half mile wide lane in each direction divided by a quarter mile wide traffic separation zone as a safety buffer, similar to a highway's median. It is like having two sets of roads, one to the east and one to the west, each allowing for traffic to travel both to the north and south.

The Coast Guard and the Vessel Traffic Service—the "VTS"—based out of Seattle control these lanes. For most relatively small boats, like the *Sea Hunt*, the tongue-in-cheek "law of gross tonnage" prevails: as a practical matter, big ships always have the right of way over a comparatively tiny pleasure craft. Typically, the VTS can help recreational boaters by providing timely radio warnings of impending conflicts with commercial traffic, but only if the small boat Captain asks for VTS assistance. The large vessels that ply Rosario Strait must comply with strict rules while operating within the shipping lanes. Each vessel must carry a device similar to an aircraft transponder to automatically transmit position, course, and speed to both the VTS headquarters in Seattle and all other similarly equipped ships in the area. This device is called an Automatic Identification System, or AIS. An AIS also broadcasts a ship's type, name, registration, and usually a photograph as well, with all information available on the Internet.

Recreational boaters are not obligated to participate in the vessel traffic scheme. The rules can be intimidating to pleasure boaters, and the regulation book is long and difficult to read. It is also expensive to purchase the required navigation gear needed to share information with commercial shipping. Fortunately, it is only modestly inconvenient for a boater in the frequent late summer fog to cross the shipping lanes safely, so long as he or she has an operating radar to locate and avoid other vessels. Still, stories circulate of hapless pleasure boaters venturing into the shipping lanes and colliding with a ship or a barge, usually out of tragic ignorance.

By sunset, we had almost completed our survey of the shallows. I took the helm from Jared and finished the final two runs in the dark. After setting a course for Deception Pass, I turned the autopilot on while Jared went forward to the V-berth to catch some sleep. Ben reviewed his smartphone for work messages, leaving me at the helm, dead tired, but unable to sit still. Our slow speed return journey was hardly relaxing. Between navigating and straining eyes into the darkness looking for floating logs

The Lost Intruder

torn loose from rafts of towed timber during winter storms, it took us well over an hour to transit the seven miles to reach Deception Pass.

The current abruptly shifted approaching the Deception Pass Bridge, causing *Sea Hunt* to accelerate. The southbound roadway at the top of the bridge, visible from the water's surface, produced an optical illusion of sideward movement as a continuous stream of dim headlights moved from left to right onto Whidbey Island. The seven-knot current gathered at the bottleneck beneath the bridge, increasing in energy until it was impossible for the small gap to accommodate all the incoming water. The torrent's outer edges were repelled by the jagged cliff face on Center Island. The rocks split the water flow into sharp reversals before thrusting it straight down more than 100-feet to the bottom. The rest of the trapped water dispersed in all directions, eventually either joining the primary inbound run or reaching a temporary equilibrium outbound and forming an eddy. The result was a stacked pile of high energy water outside Deception Pass, releasing its frustration through the formation of dozens of turbulent whirlpools, each swirling to the unseen effects of the bottom's rugged topography.

Once losing sight of the roadway under the bridge, the sudden blackness was disorienting. With the power barely advanced enough to maintain steerage, the chart plotter still indicated twelve knots due to the current on our tail. *Sea Hunt* swung in violent arcs as it was slammed side to side between whirlpools. I focused on the chart plotter and radar, struggling to keep my eyes open, unable to see the cliff faces just one hundred feet to either side in the darkness. The boat wallowed slowly, increasing its oscillations until it was pitching in rapid bow reversals. *Sea Hunt's* helm was in constant motion to avoid a potentially disastrous broach until, suddenly, we were beyond the worst of the rough water. *Sea Hunt* decelerated to seven knots as the Pass's narrow bridge channel opened to a half mile wide bay.

Several minutes later, we rounded the corner of Ben Ure Island, still moving uncomfortably fast in the gloom. Swinging the helm to starboard, I retarded the throttles and stared at the radar as we hugged the island's rocky shoreline, leaving the current behind. There had been half a dozen anchored vessels in the shallow bay leading to the marina thirteen hours earlier, but that meant little now. They could all be gone—or not—only the radar could tell us for certain. *Sea Hunt* completed the half circle turn around Ben Ure Island and motored slowly toward the marina. The adrenaline from the night transit through Deception Pass dissipated, allowing the day's fatigue to return with a vengeance. I pulled back the throttles while shaking my head, trying to stay alert.

"Ben, back me up, please," I fought to be heard over the engines' idling rumble. "I'm fading here, having trouble with my situational awareness." Ben immediately moved forward from where he had been standing silently just out of view to my left.

"Should I wake up Jared?" he asked. I shook my head.

"No, don't bother. We'll be through this before Jared's eyes can adjust to the lights. He's never run the boat at night before, anyway." I tried to expel more air with my voice and was rewarded with the deep cough of deflated lungs. The dark shape of an anchored sailboat slid by us thirty feet to port.

The radar screen started spinning slowly. I squeezed my eyes shut, then looked back at the display. The physical distraction worked to temporarily re-frame my senses; the radar screen again appeared normal. We dodged the anchored boats until *Sea Hunt* split the green and red piling lights marking the safety of the marina channel. Two minutes later, I spun the boat around in a well-practiced pirouette of opposing engine directions and backed into the brightly lit slip. It was almost midnight. I hauled my tired legs over the transom onto the dock and tied up the boat. It was only the sound of *Sea Hunt's* rough idle going silent that finally woke a bleary-eyed Jared. I sent him to the parking lot to start the car while I straightened up. Ben packed his gear in a nearby dock cart, said a quick goodbye, and headed up the walkway for the long drive home.

This was to be our last opportunity to search for the month of July, and I welcomed the break. The two days of searching highlighted the need for several essential maintenance repairs, the most pressing being the availability of clean electrical power. The sonar's multiple electronic displays, all highly sensitive to irregularities in the AC electrical current, ran continuously. *Sea Hunt's* 27-year old wiring and undersized alternators were unable to provide sufficiently stable electricity to power the multiple computer monitors. The boat's inverter, the device used to transform the DC power from *Sea Hunt's* batteries to a computer monitor friendly AC power, created a snowstorm of static on Ben's laptop. We had been forced to run the portable Honda backup generator all day to provide clean AC power. Operating the generator wasn't necessarily a problem, but it took up room on deck and required electrical cables to run to the displays inside the cabin, creating a constant trip hazard.

The long day yielded a single sonar find. About a half mile south of Colville Island during one of the last daylight runs, Ben saw what he described as a series of tiny contacts that looked like a small debris field in 200-feet of water. The scattered pieces appeared to be too small to be from an A-6 unless the jet had disintegrated on impact, a possibility I didn't even want to consider. We had combed the new sector thoroughly but had little to show for the effort. I immersed myself in solving the electrical problem, temporarily pushing the missing A-6 to the back of my mind. I honestly didn't know what to do next.

One week later, relief came in the form of a U.S. mail package. The return address on the small padded, manila envelope read, "Department of the Navy, Office of the Judge Advocate General." Taking a deep breath, I pulled the envelope's easy-open strip, immediately saw the compact disc, and felt a wave of excitement. The FOIA request had been granted. The JAG Investigation had been scanned, and the report's CD mailed to me in less than a month.

The receipt of the JAG Investigation was timely beyond concern over the summer's good weather slipping away. The break from the active search should have been good for my health, but the truth was that I couldn't take advantage of the opportunity to rest, not with my symptoms worsening so quickly. My reaction to the disease's progression was questionable at best—I pushed DBS surgery back a month until November. My reasoning was simple: I figured the few remaining months of 2014 might be my final opportunity to find 510.

But that wasn't my only reason for postponing the surgery. The neurosurgeon had explained that there was a 1% chance of life-threatening complications. This was damn good odds in Vegas but suddenly didn't sound so great when put in terms of brain surgery. Still, the physical danger of the procedure had nothing to do with the change of dates. What gave me pause was a deep-seated dread of losing a hard-fought-for awareness and inner peace.

I had discovered a degree of spiritual freedom through an honest, unfiltered understanding of myself. I could accept the looks of others, from horror to disdain. For the better part of my life, my identity had come from external sources. The recognition that my ultimate identity could only be determined by me was beyond empowering.

Now free of ideas and concepts that had dominated my thinking for a lifetime, I was emboldened to do what a few years earlier would have been unthinkable. As a local school board director, I spoke unreservedly at meetings in a nearly incoherent mumble, eyes wielding a fiery passion to be understood. There was no self-consciousness as I dragged my right leg and stumbled awkwardly through the supermarket or across town. It didn't matter. The old me had done the same things at times, but that had been posturing, attempts to pretend to have found an inner peace to keep my family strong. The search for the lost Intruder honed my focus, it hardened my resolve, turning the pretense into something real. To find the missing A-6 would be to wrench an achievement from life that the U.S. Navy had been unable to secure. It would not just be a win in the privileged class of the victim of a debilitating disease. It would be winning on anyone's terms. I never felt more alive.

The DBS surgery scared me with the thought of losing this hard-won prize, but I was also concerned that delaying the surgery might negatively affect the outcome of the procedure. My family was battling with the lost Intruder for my attention. I looked for balance, found it in a compromise, and pushed the operation back six weeks. In my mind, there was still an outside chance that the delay might cause me to miss a narrow physiological window that could impact the success of the procedure. I decided it was a risk worth taking.

A poignant Naval Aviation saying describes the sensation of tracking on an out of control path to disaster. When events drive free will, one "is letting the plane fly you." The lesson is simple. Don't allow evolving circumstance to control your actions; don't let yourself be led in a known wrong direction. Take back control of the aircraft; put the jet where it needs to be. I could control my own destiny, not forever, and not necessarily in a steady direction, but the general course was still mine for now.

Was it a choice between happiness and surgery? Maybe. By refusing to allow some obscure destiny of diagnosis to be my guide, I had wrestled back control of the aircraft. If there was ever to be peace on my death bed, then I must be responsible for my actions. I didn't want to take the chance, almost certainly greater than the 1% odds of DBS going wrong, that I would lose this positive sense of well-being forever. Yet, I also recognized the need to undergo surgery, if for no other reason than for the sake of my family. This, too, was a part of "flying the plane." It is not just about being in relative control of whether one lives, but in how one lives.

It was not a bad place to be after 52 years of life. Somehow, I knew it was possible for the surgery to improve my physical situation without losing the lessons from my first descent into darkness. But it would be one hell of a challenge. It was intuitively imperative that my sense of self-determined identity remain intact. It meant everything that I keep flying the plane.

I looked down at the package and considered the CD. It was like opening a time capsule, and the seductive promise of easy answers beckoned. Warning signs shot through my brain with electric speed. If the years had taught me anything, it was that few noble pursuits reach a satisfactory conclusion due to easy answers. What was the end-goal of my search? When I eventually found the lost Intruder, would it be enough?

Eleven

...CAULKED NERVE HOLDS MUSCLE TO STEEL

July 2014

I struggled to grip the CD case in the small envelope. Patience is not the type of skill that is willingly practiced. In fact, most people don't consider patience a skill at all: a virtue, perhaps, but not as a means to an end. Skill requires practice. Most people, including me, only put patience into practice when the exigencies of actual circumstance dictate. In recent months, my opportunities to exercise patience had abounded. The body has amazing coping mechanisms. It prepares for events that the mind subconsciously must know are coming, but have not yet been brought forward to a conscious level of cognitive recognition. On the fifth attempt, I was finally able to maintain a tenuous hold of the JAG Investigation CD case and pull it out from the padded envelope. After placing the CD in the laptop to boot up, I began reading the JAG office's reply letter.

The letter ominously stating that my FOIA request had been "granted in part and denied in part." The denial turned out to be nothing. The only change to the report was the de-identification of names in the investigation. As a practical matter, this translated to a simple line through most names in the report with a thick black marker, a non-issue for my purposes. I opened the file entitled "*A-6 Investigation Redacted*" and scanned the disc's contents.

The Lost Intruder

Two hundred fifteen pages of facts, logic, some speculation, and very few black ink obliterated names. The report had been submitted on its due date to arrive at the first endorsee, the N.A.S. Whidbey Air Wing Commander, on December 6, 1989, exactly one month after the mishap. The first section consisted of endorsements by officers of increasing rank as the document went up five more levels in the chain of command, including the Commander of the U.S. Pacific Fleet, before arriving at the office of the Judge Advocate General on June 26, 1990. It had taken over six and a half months for the report to be reviewed by the chain of command after being investigated, compiled, and written in thirty days.

A thought crossed my mind. I pulled out the *Salvor* report and scanned the front page for a date. The *Salvor* report had been submitted to the Naval Sea Systems Command on March 25, 1990, a full month after the search for 510 had been called off. The *Salvor's* search effort, therefore, did not have access to the JAG Investigation, or presumably the Flight Mishap Report, which was likely to have been written and submitted on a parallel timeline. It was unlikely that anyone other than the endorsees was permitted to see either document until they had run up the chain of command.

I tried to put myself in the shoes of the *Salvor's* commanding officer. He probably spent little time attempting to discover the exact crash point, opting instead to survey as much of the ocean bottom as possible. It made sense, the professional diving officer playing to his area of expertise instead of wasting time trying to navigate Naval Aviation's obscure acronyms and procedures. We were taking the opposite approach by leveraging my past A-6 flying experience to narrow down the search area.

But the *Salvor* still had to start somewhere; they must have relied on some basic assumptions to center their initial search pattern. Where would the Navy salvage crew have gotten this information? If posed this question in 1989, I would have immediately pointed to air traffic control. The air traffic control sailors worked the local radars, keeping track of all aircraft in their jurisdiction. They would have had hard radar data on the jet's exact position at ejection and possibly shortly before the crash. The radar track of the two A-6s and the statements of the tower personnel were both listed in the JAG Investigation's index. Even better, the radio transcripts were included, which tied events to real-time to the second. The *Salvor* might have been able to get some of this information in bits and pieces, but probably not in a neat, organized format. It would have been hard for the Navy divers to know which information was useful.

I decided to plot the radar returns of the two A-6s to recreate the scenario in detail. The radar data was laid out in a cramped, tabular format. The first column was annotated "Plot symbol," with either an "A" or an asterisk listed in the entries below. This must represent the altitude reporting function of the transponder, I reasoned. All radar contacts without transponder associated altitude readouts were represented by an asterisk instead of the letter A in the plot symbol column. The second column heading was "Code." The 116 entries under the code column were all the same, the four-digit number 4667, presumably the flight's assigned transponder code, although there were about a dozen blank spaces toward the bottom. I figured that these empty entries might correspond to radar hits of an aircraft without an active transponder. The radar returns of a target without a transponder code were called "skin-paints." Each skin-paint would be blank in the "code" and "altitude" columns.

The remaining columns contained real times in 12-second increments and the associated aircraft positions in latitude and longitude coordinates. The data appeared to be a confusing jumble of numbers, letters, and symbols, but once plotted on a chart, the individual entries marked 510's position until the jet had descended through an altitude too low for a radar skin-paint to register. This would not be theory or conjecture or eyewitness testimony. The plotted data would be hard fact.

I flipped through the report until reaching the flight's radio transcripts. All communications between the two Intruders, the tower, and approach control, including the exact times for each transmission, were chronologically listed. The report contained written statements and interviews of each witness to the accident as well. There was also an inventory of the floating debris picked up by the Coast Guard Cutter *Point Doran* shortly after the crash, as well as the statements of the search and rescue helicopter's crew. The JAG Investigation was rounded out with about a hundred pages of 510's past maintenance records. The wide assortment and detail of the report's enclosures were incredible to see after developing a theory based on educated guesses. I sat down to read the JAG Investigation from start to finish.

Once confronted with the sheer volume of available data, my initial theory seemed childishly simple. Another thought occurred to me. After hours of observing Ben, a professional sonar operator, I had developed an appreciation for both the required skill and the impact of outside forces on attempts to survey the ocean bottom. It was understandable why weather and currents could quickly render even multiple sonar runs over the same spot worthless. The Navy's search lasted a total of nine weeks in the dead of a Washington winter: it might have been impossible to thoroughly explore the entire high probability grid in the time allotted.

Had the *Salvor's* intent been to conduct just a single side-scan pass over vast swaths of ocean bottom? It was also curious to note that nowhere in the *Salvor* report were the words "slack current" mentioned. The prospect of a professional salvage crew not adjusting their schedule to the fast tidal flow made no sense to me.

It was a safe bet that 510 crashed in an area of significant current, leaving me to speculate that perhaps our greatest advantage was the time of year. The *Salvor's* effort had involved multiple ships all searching on the same limited number of winter days. Inclement weather would have negatively affected each vessel relatively equally, compounding unreliable data on the days of adverse meteorological conditions. Our search was being conducted in the far calmer seasons of spring and summer. Perhaps more importantly, our lone boat strategy was dependent on a sound theory to identify key high-probability sectors to survey, if need be, over multiple days. With the JAG Investigation now in hand, this seemed to be a winning strategy. But it all depended on tight, accurate research.

I leafed through the report's copy of the squadron flight schedule, weather briefing, and squadron duty officer's account of events, none of which offered any new information. I read Commander Starling's statement and interview with the Investigating Officer. Except for Starling's estimate of the jet's impact distance from shore, his account seemed to align well with Chris Eagle's testimony.

Next were the 502 aircrew statements of Ray Roberts and Rivers Cleveland. Rivers Cleveland's testimony yielded nothing new. Then I turned to Ray Roberts's account. My pulse began to race. Ray had aggressively maneuvered 502 in a break turn to the right to avoid hitting Chris Eagle after he ejected. Once clear of Eagle, Ray Roberts had leveled his wings to regain sight of the now crewless 510. From that point on, Ray Roberts had watched 510 almost uninterrupted until water impact. Chris Eagle was not the only eyewitness to the crash; Ray Roberts had seen it too. Ray's statement was short and terse:

"...bombardier/navigator initiated ejection, followed shortly by the pilot. We broke hard, up and right. We saw two good chutes. Aircraft 510 rolled to the right and entered (the) water in a nose down wings level attitude. Ejection altitude was 3500 feet."

The statement answered two critical questions: 510 had flown mostly wings level, straight ahead, and upright until water impact, and the ejection altitude was only 3,500 feet, not the 6,000 feet I had assumed earlier. I read on.

"We became on scene commander and followed the chutes in the descent. One crewman deployed his life raft immediately. The second crewman waited. Both (parachuted) into the water with rafts deployed."

Ray must have been in an excellent viewing position to see Denby Starling and Chris Eagle each deploy their one-man survival raft while parachuting down to the water. Rivers Cleveland, to the contrary, saw almost none of Starling and Eagle's parachute descent, but this was not surprising. With the A-6's side-by-side seating arrangement, it was common for only one crewmember to be able to see anything of note while turning. Ray Roberts had positioned 502 so that he would have an unobstructed view of the crash from the left side of the jet. If given a choice, A-6 pilots typically turned to the left. This way they could maintain a lookout for other aircraft, or in this case parachutes, in the direction they were turning. I would have done the same thing. I finished reading Ray's statement.

"Aircraft 502 then descended to about 250 feet to check the condition of 510's crew...Radio contact (made) with Commander Starling and

Lieutenant Eagle. Aircraft 502 climbed to 1,000 feet and directed helo to southern most survivor. Approximately five to seven minutes later both of 510's aircrew were aboard helo and en route Naval Hospital. Aircraft 502 returned to base without incident."

The next section of the report was the Investigating Officer's summary of his interview with Ray Roberts. About halfway down the first page, it became interesting.

"Investigating Officer: Describe the aircraft movements from the point where the ejections occurred to impact.

Ray Roberts: After (Starling radioed) that he was losing his combined (hydraulic) system, the aircraft did a wing wag to both sides and a porpoise type nose oscillation.

Investigating Officer: How much pitch did you see?

Ray Roberts: Very minor, but enough that I could see it and that the aircrew would have definitely felt that they had a controllability problem. After the ejection, the aircraft rolled right 10 to 15 degrees, and the nose fell through, beginning its descent. As it accelerated, it seemed to roll back to a wings level attitude, no more than 15-degrees nose low. At that point, we were watching the parachutes and lost track of the aircraft until just before impact.

Investigating Officer: How far off the initial heading would you say the aircraft had turned in its descent?

Ray Roberts: About 5 to 10 degrees to the right.

Investigating Officer: Describe the aircraft impact.

Ray Roberts: It appeared to impact in the same attitude I had previously seen it, nose low and wings level.

Investigating Officer: Along the same track as before?

Ray Roberts: Yes, I saw the (ejection) seats hit (the water) and then the aircraft essentially along the same track.

Investigating Officer: About how far off the islands did it impact?

Ray Roberts: Approximately ¾ to 1 mile.

Investigating Officer: Do you feel confident in what you describe as the impact attitude (of the aircraft)?

Ray Roberts: Yes, fairly confident. It was a snapshot look at the impact because I was also trying to watch the chutes. There was no explosion visible on impact."

The interview ended.

Before the ejection, 502's support was limited to providing a set of outside eyes for an exterior inspection of 510. After the ejection, 502's role automatically switched to that of on-scene-commander, directing the search and rescue helicopter to the survivors and acting as a communications relay. I turned back to the air traffic control section. It was dated 13 November 1989. The cover page was signed by the civilian Air Traffic Manager for the greater Seattle airspace, a designated sector of the sky that included N.A.S. Whidbey Island. The memorandum read:

"Attached are copies of Seattle Center records requested by you in support of your investigation of the accident involving (159572) on November 6, 1989. Three separate (radar plotted tracks) are enclosed, together they provide a full picture of the flight. Our computer was also able to calculate distance measurements. The last radar depiction showed the aircraft 4.8 nautical miles from N.A.S. Whidbey Island (at an altitude of) 1,100 feet."

Five-ten's final radar return had shown the A-6 to be 4.8 nautical miles from the control tower descending through 1,100 feet. After that point, the Intruder was too low to be tracked. The position and altitude at the time of the aircrew's ejection were the starting parameters of what Ray Roberts's statement indicated was a relatively steady state flight straight ahead into the water. I could now plot the latitude, longitude, and altitude of the ejection point in three dimensions, as well as the same for 510's last known position.

Drawing a straight line from the ejection point to approach control's last radar fix would indicate 510's exact course before water impact. The slope of the line from the ejection point altitude, 3,500 feet, to the height above water at the last radar fix of 1,100 feet would point vertically to the crash site. Determining the distance flown in the jet's descent from 1,100 feet to the water's surface was a simple extension of the existing flight path and descent rate. An extremely precise latitude and longitude at 510's crash point could then be pulled right off the chart.

Looking at it another way, two points define a line. Five-ten's position at 3,500 feet and 1,100 feet were my two points. The extension of the line formed by these points in the three dimensions of flight represented 510's final course and altitudes before hitting the water. I drew a right triangle cross section of the Intruder's vertical flight path. The distance of the vertical leg of the triangle was 3,500 feet. The angle downward to 1,100 feet, a mid-leg point on the triangle's hypotenuse, could be measured. The angle at the base of the triangle was 90 degrees. This was enough information to figure out the length of the remaining leg which would give the distance from the ejection point to the crash site. It was basic geometry.

What made the information shocking was that it was all available on November 13, 1989, well before *Salvor* even arrived in Washington waters. I started skimming through the air traffic control section of the report and was not surprised to find my work already completed. Air traffic control had extrapolated 510's water impact point on a handwritten chart. The *Salvor* crew would have had to be fools not to use this data.

I pulled out the *Salvor* report and leafed through the pages. Now armed with a chronological radar log of 510's flight path, the *Salvor* report became even more confusing. I quickly skimmed through the *Salvor* report's timeline. The *Salvor* did not even arrive at Whidbey Island until February 9th, 1990, over three months after the ejection. From the end of December until February, a single minesweeper searched most of the area before turning command over to the higher ranking *Salvor* Commanding Officer. The underlying assumptions of the search in the *Salvor* report were likely third-hand information at best. The minesweeper's six weeks of initial searching had set the tone for the effort; they had made the initial assumptions.

There was nothing inherently wrong with this, other than, as everyone who has ever played the game "telephone" knows, each iteration of an original story tends to stray from the facts. Soon, the actual story becomes unrecognizable. Understanding how the Navy operated in 1989, the *Salvor* report finally made sense to me. It was a third-hand account written by a ship's Skipper who arrived in the last three weeks of an active nine-week search. The latitude and longitude points used for the Navy's search were explicitly stated, but the source of this data was unnamed, possibly because by the time the report was written, nobody remembered or cared.

This inherent confusion might be why the Navy chose to use such a peculiar "ground zero" as a central reference point to draw rectangular search grids. The *Salvor* report listed a latitude and longitude for the assumed crash point but did not make it their ground zero. Instead, the Navy used the aircrew's ejection position based on a TACAN fix. Using the assumed ejection position—when the A-6 was still 3,500 feet in the air—as the basis for drawing search grids was surprising, to say the least. The ejection point was possibly miles from where the Intruder crashed. Even more baffling, on February 14, 1990, the *Salvor* moved the assumed crash point to a slightly different latitude and longitude position. The listed rationale for this change was that it was based on a "compilation of data and extrapolation into a search plan," a confounding sentence that told me exactly nothing.

But the fact remained that despite using the ejection point as the center of their search, the Navy's comprehensive effort had still positively identified every contact found within 29.3 square miles. So why didn't they find 510? It appeared that, once again, it came back to conditions of weather and currents. I reviewed the *Salvor* report's timeline and found three separate mentions of "excessive currents," some reaching a stunning four nautical miles per hour, making the search for that day impossible.

It was a losing proposition to bet against the air traffic control radar returns, but that is exactly what it appeared the Navy had done. The only logical reason to ignore the radar data was if its validity were suspect, but these radar controllers worked the same air space day in and out, they were the subject matter experts. What, if anything, could have cast doubt on the air traffic control calculated crash site data?

Twelve

...FOR UNSHAKEABLE KEN OF THE REAL

July/August 2014

I flipped back through the JAG Investigation until reaching the beginning of the air traffic control section. The first enclosure was 510's hand sketched flight route based on plotted radar fixes. It depicted a seven-mile diameter circle with Smith Island in the center. The circle's perimeter was annotated with exact times and corresponding events of importance. At the twelve o'clock position was a handwritten note, "Flaps/Slats lowered." At the nine o'clock point was scribbled, "Attempted gear blow down." Heading due north at three o'clock, with the runway off 510's right wing, the entry read, "Nose gear pump down attempted." Five-ten continued north, almost completing the circle before reaching the final one-word caption, "Ejection."

A total hydraulic failure caused the crash, but it was the inoperable landing gear blow down system that kept 510 from landing immediately upon reaching the field. If the landing gear had extended with the activation of the backup system, then there should have been enough time to get the jet safely on deck. The cause of this unrelated malfunction was never determined.

It was troubling that the *Salvor's* assumed ejection point was more than a half a mile away from air traffic control's radar fix of 510 at the time of ejection. I turned to the radio transcripts of the dialogue between the pair of Intruders and the N.A.S. Whidbey tower controller. At exactly 12:24 pm and 25 seconds, River's Cleveland transmitted, "Ejecting, ejecting." The transmission was then temporarily drowned out by the warbled distress tone of 510's automatic emergency locator beacon activating as the ejection seats left the aircraft. Rivers followed up within the next minute with several additional, brief radio communications: "Two good chutes," "One good raft," and "Seven miles on the two-eight-zero (compass radial)."

I began plotting the radar log's latitude and longitude coordinates in Google Earth in chronological order. Five-ten's transponder accentuated flight path took shape as each radar sweep was marked at twelve-second intervals. The successive radar blips gradually circled Smith Island closely to the south before forming an elongated racetrack pattern toward Lopez Island to the north, marching along in a predictable oval. Then, suddenly, the circular pattern went out the window.

The last radar plot, the one that presumably showed 510's transponder while descending through 1,100 feet, did not follow the established twelve-second pattern spacing. This critical, last radar blip did not appear until more than three minutes after the ejection. Equally confounding were fourteen skin-paints, raw radar returns that were not plotted by air traffic control anywhere. Why were all these skin-paints ignored? My confidence in the air traffic control conclusions plummeted.

I looked back at the air traffic control hand drawn flight path, comparing it to the radar fixes I had just plotted in Google Earth. The two routes through the sky were generally similar, but the air traffic control version was not marked at regular time intervals. The air traffic control spacing between radar returns ranged anywhere from 24 seconds to over 3 minutes.

Air traffic control's drawing showed a straight line depicting 510's three-minute flight without an active transponder from the ejection point to its last known fix at 1,100 feet. The only logged entries during this three-minute period were the fourteen radar skin paints. The skin-paints seemed to be randomly scattered, but it was hard to tell for certain, so I plotted them all in their ordered time sequence. Slowly, the scattered points took the shape of several offset circles.

Now I was really confused. According to Ray Roberts, 510 flew straight ahead after the crew ejected. Five-ten certainly did not fly in offset circles that terminated, unbelievably, on the same course line as at the time of ejection. Even more mystifying, it was only after three silent post-ejection minutes that the transponder code magically reappeared for a single active readout at an altitude of 1,100 feet. If this was to be believed, it meant that after the crew ejected, 510 flew for over three minutes straight ahead without an operating transponder. During this time, 510 only descended 2,400-feet, a descent rate of 800-feet per minute. This was the controlled descent rate of a normal landing.

My experience flying Intruders told me that this was impossible. Five-ten was traveling just above stall speed at 120 knots at the time of ejection. If it had flown straight ahead for three minutes, never mind the additional time it would take to reach the water from 1,100 feet, then it would have traveled six nautical miles down range, which would have put 510 over the northern tip of Lopez Island when it crashed. Impossible. That was not consistent with Ray Roberts's eyewitness account. Nor was it compatible with the timetable from the radio transcripts, which indicated that both crewmembers were in the water after a 2 minute and 15-second parachute descent. By all accounts, 510 crashed before Starling and Eagle reached the water.

Another inconsistency bothered me. Why would 510's transponder suddenly start working again after three minutes of silence? The A-6 transponder control panel was located on the center console between the pilot and the bombardier/navigator. Could the transponder have survived the blasts from the two ejection seat rocket motors? Doubtful. What would have caused it to start operating again? None of it made any sense. What, then, was air traffic control looking at after a three-minute dead period, when the transponder identification code of 4667—the single identification code assigned to both 510 and 502—miraculously reappeared?

Then it hit me. Where was the wingman? What happened to 502 after the ejection?

"Holy shit!" I said out loud. Suddenly, it all made sense. Air traffic control was not looking at 510 during the three minutes after ejection—the fourteen skin paints and the final 1,100-foot transponder hit were all from 502. After ejection, air traffic control never saw 510 on radar again.

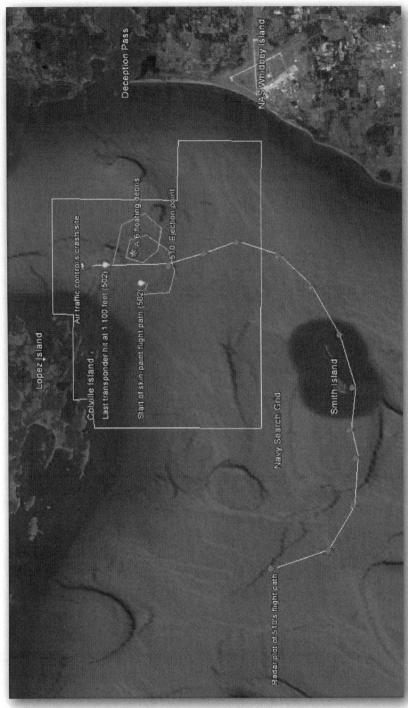

Radar log of the flight paths of 510 and 502 (adapted by the author from Google Earth screen shot).

The Lost Intruder

Five-ten automatically took the formation's lead position when the jet's first hydraulic system failed. As the lead was passed to 510, Chris Eagle moved 510's transponder switch to "on." Rivers Cleveland did the opposite, turning 502's transponder to "standby." Both jets would have taken off from N.A.S. Whidbey with the same code of 4667 dialed into their respective transponders, but only the jet flying in the lead position would turn it on.

When 510's crew ejected, Ray Roberts was forced to make an evasive break turn to avoid hitting Chris Eagle as he rocketed toward 502 in his ejection seat. A break turn is designed to "break" a radar missile lock. For the next several radar sweeps, 502 disappeared from the radar scopes, not even registering a skin-paint. When Ray regained sight of both Starling and Eagle, now swinging in their parachutes, he started a left-hand orbit above their position. Standard procedure dictated that the wingman, now without a lead to operate the transponder for both aircraft, would turn on his transponder to highlight their position to air traffic control. Given the urgency of the ejection, a last-ditch effort to survive, activating 502's transponder was not a priority. It probably took Rivers Cleveland three minutes—about the time for Starling and Eagle to reach the water and get into their rafts—to remember to turn on 502's transponder.

Once 510's crew was in their rafts, 502 rapidly descended to 250 feet. Ray Roberts did this to check on the physical condition of the downed crew. Rivers Cleveland got around to turning on 502's transponder as they shot passed 1,100 feet, leading radar controllers to believe that 510's transponder started working again for a single radar sweep before crashing. By the next radar cycle, 502 was virtually invisible, flying below air traffic control's radar coverage at just 250 feet above the water.

When 502 climbed back up to an altitude where it was visible to radar again, air traffic control had stopped collecting radar log entries. The air traffic controllers never saw the aircrew statements, and in the absence of other evidence, the controller's interpretation of events made sense. The last transponder hit descending through 1,100 feet was never correctly attributed to Roberts and Cleveland in 502. This meant that the most critical evidence in the JAG Investigation regarding the crash site was in gross error. Even if the *Salvor* did concentrate their search based on air traffic control's conclusions, they would be looking in the wrong place, just as Chris Eagle had said.

The Google Earth chart was getting cluttered, so I decided to color code the wingman's flight path light blue to clearly differentiate it from lead's. Once these fourteen points were discounted, it became apparent that 510's course at ejection had to be due northwest.

I scrolled past the entire air traffic control section until reaching the list of debris found by the Coast Guard Cutter *Point Doran* eighty minutes after the ejection. I plotted the latitude and longitude of the floating debris field on the Google Earth chart, minimized the screen display, and opened an online tide and current program I often used for dive planning.

The tidal software confirmed that the ejection occurred almost exactly at the high-tide slack current. I went back to the tower radio transcripts. Rivers Cleveland had transmitted the low altitude winds to assist the search and rescue helicopter enter a hover. They had been coming out of the south/southwest at 16 knots. Therefore, the wind would have pushed the floating A-6 debris to the north/northeast for eighty minutes before the *Point Doran* retrieved the wreckage. Unlike the current, the wind would move floating debris decidedly slower than its own speed. The actual time and duration of slack tide varied widely in Rosario Strait: it could be mere minutes or over an hour long. The only way to accurately gauge the zero-current time for certain was to go on site. But I didn't know exactly where the crash occurred; I only knew where the Coast Guard Cutter had found the floating aircraft debris. The computer modeled current for the general vicinity of the crash would have to be close enough for now.

Between the current and wind, I estimated that the A-6 floating debris moved less than a mile to the northeast before being picked up. Working the problem backward, I drew an arc a mile to the southwest of the debris field on Google Earth. I then pushed the search area a bit north to account for the aircrew distance estimates from Lopez Island. I placed a one nautical mile diameter circle around this new crash point.

The Lost Intruder

My revised water impact site was at the edge of the *Salvor's* high probability search grids. Air traffic control's estimated crash site was a mile and a half further to the northeast. The new impact point was a bit over a mile equidistant from Colville and Lopez Islands. I cross checked the nautical chart on the wall and saw that the depth ranged between two and three hundred feet. Narrowing down the crash zone any further was a guessing game. We would need Ben's deep-water sonar if we were to have any chance of finding the missing A-6. But I was confident that we had whittled down the search area to about one square mile.

Excited, I emailed the other team members the good news, including several Google Earth screen shots to help them understand my reasoning. It was complicated and difficult to explain, but at the end of the day, we had a new theory based almost entirely on fact. Good news followed. In a feat of logistics magic, Ben had arranged to borrow an industrial strength winch from a professional diving salvage company in Seattle.

But we would have to wait a week to resume the search. Laurie and I were leaving on the boat for a long-planned vacation to the San Juan Islands until August 10. We decided to load the 700-pound winch and its gasoline motor power unit onto *Sea Hunt's* small aft deck on Tuesday, August 12. I drove the ten minutes to the marina and found the local marine salvage/repair shop owner, Captain John Ayedelotte, in his boat yard. We scheduled the free-of-charge on-load of the winch for the 5:00 pm high tide. Our search would resume the next day.

When Laurie and I returned, however, it was with a failed port alternator. Several phone calls Monday morning confirmed that replacement parts were no longer available. We absolutely needed that alternator; the boat's electronic gear could only operate for a short period on batteries alone. The clock was ticking on the loan of the winch. The end of the summer's favorable weather and the date of brain surgery were fast approaching. I had two days to find, buy, and install a new alternator.

For most people of modest mechanical inclination, this would have posed no great challenge. But Parkinson's disease changes everything. It meant numerous obstacles to acquiring the alternator in time, never mind installing the unit. I had grown to relish this sort of test and refused to ask for help. I would succeed or fail under the full weight of the disease.

I began making calls looking for a store able to custom fashion an alternator on the spot, eventually finding a shop in Bellingham that advertised custom built marine alternators in just three hours. There was just one problem: Bellingham was over an hour away, further than I usually felt capable of safely driving. To avoid having my right leg and arm freeze up while on Interstate 5, I took my medications slightly early. It would end up taking over three hours to get to my destination.

Barely thirty minutes into the drive it became evident that I had badly misjudged my medication and needed to get off the road quickly. Fortunately, there was a McDonald's at the intersection of State Route 20 and the Interstate. By the time I pulled into the parking lot, I was twisting in a painful full-body contortion. Steering mostly with my knees, I pulled into the first open parking spot.

Getting my medication dosages right had become an increasingly impossible balancing act. Waiting as little as thirty minutes too long, or taking too small a dose, would result in debilitating dystonia as my right ankle and wrist curled in painful muscle contractions. If this happened, I could only travel short distances using the two collapsible canes kept in the backpack that accompanied me everywhere. Any distance greater than a few hundred feet required crutches. Taking the meds too early could mean a temporary overdose and the full-body sways and writhing of dyskinesia.

It was a beautiful August day, but it felt as though I was stumbling through a dense fog. The excessive medication had come to subtly affect my cognition, making the simplest decision a confusing chore. It was as if my thoughts were being filtered through a dark blur. Once reaching my brain, a slow-motion understanding of a problem would evolve, but translating comprehension into action could take frustratingly long.

Opening the McDonald's door released a cacophony of children's voices coming from the indoor playland. I ignored quickly averted stares as mothers steered their children away from me. Staggering to the order counter, I knew that I needed to eat to keep up my strength, but was also aware that I must regain control quickly if there was any chance of getting a working alternator that same day.

The Lost Intruder

The kid behind the McDonald's counter could barely contain his disdain for what he probably took for an early afternoon drunk. Fighting for breath, I strained my voice to order while trying to stand well apart from the other patrons to avoid presenting a threatening appearance. I watched the slow transformation cross the server's face as he realized that there was something wrong with my health. It was a familiar metamorphosis. Virtually everybody in the McDonald's noticed my jerky sway at the counter. Probably two-thirds of the spectators eventually concluded that I had some sort of illness and was not a threat. Those lacking the imagination or compassion to allow for such a possibility were discounted from my consciousness in the familiar ritual of getting through the day.

I placed the bag with three cheeseburgers and fries on a nearby booth table and sat down. After a brief struggle, I finally managed to pull my cell phone from my short's side pocket to look at the time—12:15. Any vestige of self-consciousness was abruptly masked by a mounting excitement.

Overcoming unexpected challenges had become a source of salvation to my spirit. First learned in Navy "SERE"—Survival, Evasion, Resistance, Escape—School practicing techniques to survive as a prisoner of war, the concept of small victories had helped me through hard times before. But none had been on the scale of my current battle. The idea of the smallest victory—walking with canes instead of lying down, driving beyond my usual thirty-minute range—empowered when the only other option was to succumb to a dismal frustration. I'd learned to seek out these otherwise un-noteworthy trials, even learned to devise new tests, just to experience the prospect of a small victory.

I ate slowly. Small bites, extra chewing, and patience were essential to the successful navigation of a meal. Each swallow was initiated with a bow of the head to avoid aspirating food or drink. Attempting to swallow without a concentrated effort meant racking coughs now and the threat of pneumonia shortly. I finished the first burger and thought back to recently self-manufactured tests. One stood out. It had happened in June, soon after the sonar contact identification dives. During a break from working on *Sea Hunt*, I found myself bored and looking for a challenge.

I had grown accustomed to swimming in the marina's 48-degree water. The shock of the cold would temporarily jolt me out of dyskinesia, allowing me to function with fine motor skills for short periods. After spending forty minutes trying to insert a single screw, I might jump off the dock, tread water for a few moments, climb out, dry off, and quickly manipulate the screw into its hole. The positive effect only lasted five to ten minutes, but a lot can be done with relatively steady hands in that time. It was becoming a habit to use the frigid water as a diversion, both to finish the task at hand and to escape boredom if unable to drive.

Looking across the bay from *Sea Hunt's* slip in the marina toward the Deception Pass State Park, a plan started to gel. The beach on the opposite side of the bay was about a quarter mile away, but to be sure, I verified the distance on the boat's radar. The test slowly took shape in my mind. A quarter mile swim across the chilly bay, then a hike up the embankment to the Park's rugged, backwoods trail, and finally a return hike-jog of about two and a half miles back to the marina. I was under no illusions as to the physical hurdles that would likely be encountered. A critical aspect of these invented trials was not simply to test physical limitations, but to explore creative ways to work around problems as they came up. I'd found that this helped me to devise strategies when dealing with the disease's newly introduced obstacles. But things could get dicey so I would plan the hell out of these evolutions.

My greatest concern would be encountering a bout of dystonia during the hike back. If unable to walk, crawling through the woods was a final option, but pain for its own sake was not the goal. Instead, I opted to take my next dose of Levodopa early, knowing full well that this would increase my dyskinesia, guaranteeing a weaving path in my limping-shuffle return to the marina. But it would leave me with no doubt as to my ability to continue generally moving forward. It was basic risk management: better to err on the side of too much involuntary motion than to risk becoming immobile alone in the woods. I knew that the swim was easily within a slow and steady range. The cold would help me focus.

I fashioned a rubber sling around my neck to hold the sandals needed for the hike. I then put a water bottle in each of the sandal foot pockets before snugging down the Velcro straps. I looped my watch band around one of the sandal's straps; wearing the watch was too painful to my ultra-sensitive wrists, but I needed to bring it for exact drug dosage timing. Finally, I jammed a small waterproof pill container of medications into a Velcro secured pocket before slipping quietly into the water wearing only shorts, not wanting to delay the swim's progress with a question and answer session with a passerby. I began a steady breaststroke across the bay.

Perhaps due to a current, the swim took longer than expected, and it wasn't until forty minutes later that I planted my feet in the thick mud on the opposite side of the bay. The hot sun did little at first to take the edge off my hypothermic shivers. Then, on my second step toward the beach, my foot recoiled in pain. Reaching down into the dark water, I carefully surveyed the bottom until finding a rock covered with jagged barnacles the size of my fingertips. Raising my eyes, I took a good look at the beach and immediately noticed two things. The low tide had created a 75-foot dry buffer along the embankment, and the beach was filled with small rocks, each covered with razor-sharp barnacles.

My mind started to fog as I slowly stepped forward, at first ignoring the cuts to my feet. I needed to get out of the freezing water before the situation deteriorated further. Stopping on a two-square foot island of barnacle-free mud, I pulled the rubber sling from around my neck and took stock of the sandals. One of the water bottles had worked free during the swim, and the Velcro strap that crossed the right sandal's toes was out of its plastic retaining link. I managed to unhook the watch and stuff it into my pocket. With violently shaking hands, I took a small sip of water from the second bottle before inadvertently flinging it across the beach.

I considered swimming back but decided it was too dangerous while hypothermic. Patiently attempting to control my hands, I dropped the sandal, tried to pick it up, and promptly fell on my back. The barnacles tore at my skin, but I got a hold of the sandal and stood. I tried to insert the strap again. I dropped it again. This went on for 45 minutes. I know this because the only thing I returned with besides the sandals and shorts was my watch; I ended up losing the small pill bottle as well. Finally, I secured the Velcro sandal strap back through its plastic retaining loop. It would have been impossible to exit the beach without wearing sandals, not without shredding my feet completely. By the time I reached the embankment, I had fallen a half a dozen times, and my back and legs were covered in an ugly layer of blood, mud, and thatched seaweed.

The first portion of the path through the woods wound up a steep trail to a cliff face. Sweating profusely now, my body screamed for water. In addition to physical exertion, drug-induced dry mouth made my thirst too painful to ignore. I kept up a brisk pace, dragging my right leg along the trail until reaching an opening in the trees. I was surprised to see two middle-aged women hikers enjoying the view while drinking iced coffee out of covered Starbuck's cups.

"Excuse me," I said, forcing the air out of my lungs, "I didn't mean to surprise you. I've got Parkinson's disease and am having some trouble." I couldn't tell if my voice was intelligible.

"Do you have any water?" I hated asking for help, but my thirst was unbearable. In retrospect, this was the first time I'd asked anything of a stranger due to the disease.

"No, we don't." Snapped the lady closest to me. I could tell they were tense but didn't know if it was irritation at my intrusion or a false confidence due to feeling threatened.

I took a step back. "I know this sounds strange, but can I have a sip of your coffee?" It struck me that they must have understood my first question.

The same women replied more forcibly as if I was on the verge of attacking her, "No!"

At first, I was shocked, never having been denied a simple favor when in genuine need. Then, for a moment, I got angry; black-rage angry. But I said nothing. I looked at the two for another second, then turned to continue down the path.

I attempted to raise my voice: "Have a great day."

My words were not sarcastic or angry; they were strained and painful. I tried to forget about the two women, but their response bothered me for weeks to come. All they could see was a threat; all they could consider was their own well-being. I vowed not to ask for or accept help for the remainder of the return to the marina. This was not supposed to be easy.

It took an hour of stumbling along the narrow trail before reaching the road. Once in the marina parking lot, I waved awkwardly to several friends sitting at a nearby picnic table. They got me a bottle of water, helped clean up my legs and back, and applied antibiotics. It looked like I'd just been flogged.

Parkinson's loves irony. As I turned the dock corner to my boat to fetch dry clothes, the disease left my body for twenty minutes of uninterrupted "on" time. Temporarily forgetting the coffee-drinking women, I rode the mental high of the small victory and drank a beer.

It was, without a doubt, one of the best days experienced in a long while. The fuzzy memories got me through many future moments of fleeting self-pity. They make me smile to this day. It was worth every piece of shredded skin; it was worth the hurt inflicted by the two ladies.

I took another careful bite of my McDonald's hamburger as I tried to patiently wait for my overmedication to subside. In the disease's typical mocking style, the symptoms inexplicably lingered.

Thirteen

LUNGS THIRST FOR THE ROUND TO BE OVER...

August 2014

Sitting in McDonald's, I strained at the invisible leash of responsibility that forced me to pull over. I'd learned to curb my impatience at such moments by observing the reactions of others to my erratic movements. Although mostly benign, there were still the occasional under-the-breath asides of "meth-head," or "tweaker," but the few people who felt some unassigned obligation to publicly denounce a person they had never met didn't bother me. They were the easiest to ignore.

Everyone—workers on lunch break, old ladies, scampering kids—detoured to beyond an invisible buffer around my seat until forty minutes after my arrival when, finally, I could safely drive. The rest of the day was tough but manageable, and I got home four hours after leaving McDonald's with a customized, more powerful alternator than before. No sooner had I pulled into the driveway then my smartphone started to vibrate. It was an email from Ben.

Tuesday, August 12, 2014

Peter,

I am planning to work with the salvage company here in Seattle on the winch tomorrow morning at 10:00 am. As soon as we get this loaded, I will start heading north toward you. My thought is that we would shoot for the evening high tide to load the winch and power unit.

Hopefully, everything goes smoothly, and we will be able to head out Wednesday morning for sonar scanning. Is it okay if I crash on the boat Tuesday evening? I also have Thursday available to scan.

Does this work?

Ben

I replied, "Yes to all."

It wasn't until noon the next day that I got to work in *Sea Hunt's* bilge. It took one hour to complete 90% of the alternator install, and another four to properly adjust the tensioning belt and insert shims to eliminate a wobble. I was making the last correction when the thud of feet striking planking reverberated down the dock.

Snaking my way out of the bilge, I stood and gave Ben a nod of acknowledgment. It was critical that I finish the electrical work before the winch and power unit were craned aboard, pinning the engine hatches beneath their 700-pound weight. Normally, there was no need to open the hatches while underway. There was a long list of otherwise minor problems, however, that if not addressed immediately could prove disastrous: the hydraulic steering could spring a leak, a fuel filter might clog, and on and on. Seeing no good options, I chalked it up as a required risk. I didn't think about the matter again until the winch was craned off two and a half weeks later.

Ben climbed aboard as I entered the cabin to start the engines. A minute later, we maneuvered *Sea Hunt* through the narrow gap between the fuel dock and the parking lot, sliding gently alongside the unforgiving steel bulkhead. Ben was securing the last temporary mooring line to the bulkhead's stout handrail when I heard a loud rumble. Smoke belched from the parking lot across the street as Captain John Ayedelotte started his mobile crane. Ben scrambled over the bulkhead rail and jogged to where the winch and power unit filled the bed of his pickup truck. He repositioned the pickup closer to the boat, arriving at the bulkhead just as the crane came to a standstill alongside him.

My jaw dropped—the winch was massive. Mounted on a boxed frame of four-inch steel angle iron, the winch had a three-foot diameter spool that looked like it could hold 10,000 feet of cable. Sitting next to the winch was the power unit connected to a twenty-foot hydraulic service line wrapped in a protective burlap sheath. I cringed at the thought of the rigid edges bouncing on *Sea Hunt's* fiberglass deck, and raced across the street to scavenge bracing lumber while Captain John and Ben rigged the crane's harness. Loaded down with an assortment of scrap boards, I got back just in time to see Ben disappear below the bulkhead rail as he climbed onto the boat. The winch was hanging precariously from the crane's boom, slung out over *Sea Hunt's* aft deck. Captain John cranked the crane's steering wheel to the left and inched the giant spool further out over the water.

We had barely finished laying down a protective wood bed when the crane engine revved loudly, and the winch started to lower. We jostled the winch, and then the power unit, onto two-by-fours as each reached the deck. *Sea Hunt* protested the additional weight with a waterline squat of three inches. Satisfied with the positioning, Ben secured the winch and power unit to the side-rail cleats with come-along straps. There was enough room to walk around the components on the aft deck, but just barely, and several hydraulic fittings hung precariously over the open walkway gaps. I grabbed a stack of disposable oil absorbent pads, dropping them on the deck just inside the cabin for ready access when needed. Most of my experience with hydraulics came from flying A-6s, and one lesson learned was that hydraulic fittings on a well-used piece of machinery tended to leak.

It was 7:00 pm and Ben still had work to complete from his day job as a management consultant. He left for dinner in town with laptop in hand while I straightened up scattered tools and moved the boat back to its slip. It was going to be a long week.

Dawn was breaking when I pulled into the marina parking lot the next morning. A heavy dew hung in the air, creating wisps of fog spiraling up from the bay. I felt a profound unease. Fog regularly blankets Rosario Strait in August, vying with the sun for daily dominance with unpredictable results. When the fog wins, it typically spans Rosario Strait without interruption from Lopez Island to Deception Pass.

Jason Ayedelotte hands Ben Griner bracing lumber as Captain John Ayedelotte prepares to lower the winch onto *Sea Hunt's* aft deck (photo courtesy of Ben Griner).

The prospect of a full day of sonar searching in near-zero visibility concerned me, but the winch was aboard, Ben had taken off work, *Sea Hunt* was ready to go, and the summer was flying by. I tried to tell myself that I could handle the fog, and maybe I could for an hour or two in the morning, but all afternoon? I pushed the thought from my mind.

I was finishing the last of the boat preparations when our deckhand for the day, Rob May, turned the dock corner. Rob has lived on Whidbey almost all his life and is one of those people who seems to know everybody. He immediately started asking Ben questions, discovered their shared background as firefighters and paramedics, and revisited the topic throughout the day to keep the conversation going. Rob's capable presence and easy going manner were reassuring. Feeling marginally more confident, I climbed back aboard and started the engines.

I left the diesels to warm up, stepping carefully over the stack of sonar gear scattered across the cabin floor. I eased my legs slowly over the side to avoid knocking my ankles against the winch's steel spool. Moments later, the lines were cast off and the boat was moving forward as the windshield wipers swept away the overnight condensation. The channel was empty.

In less than five minutes we were turning the half circle around Ben Ure Island, heading west toward Deception Pass under a brightening sky.

I dropped my feet to the deck for balance against *Sea Hunt's* side to side jolts as she broke through the whirlpools under the bridge. Six months into the project, this would be our first side-scan day with high confidence in our research. But there were still other obstacles. The weather could easily deteriorate into persistent fog for the remainder of August and into September, and I worried about how much longer Ben's work schedule could accommodate the project's needs. My deteriorating physical situation was also troubling; the long summer was breaking me down, more so than I realized.

Ben started letting out the transducer cable the moment we reached deep water, working the winch with Rob until confident in his abilities. The next two hours were encouragingly productive. The current was negligible and the flat bottom made the smallest of contacts hard to miss. But as the sun warmed the water's surface, it became apparent that the fog was coming.

I glanced at the Internet Vessel Traffic website on the laptop sitting on the bench seat, taking note that the shipping lanes were clear of traffic, at least for now. The Vessel Traffic freeware program should have tipped us off to the positions, courses, and speeds of all shipping in the area, but I remained wary. We had discovered that the application was occasionally slow to update, creating a false sense of security. The delay could last just long enough that a relatively fast-moving tug or commercial fishing vessel might go unseen until it was too close to avoid. In clear skies, the time-lag was no big deal, but the sudden appearance of a large boat on radar close-aboard became vitally important with the prospect of fog.

Outfitting *Sea Hunt* with a commercial grade Automatic Identification System, an AIS, that would highlight our position to the Vessel Traffic Service would take at least several days to order, register, and install. With the winch now finally aboard, it seemed foolish to wait. Without an AIS, radar was our only dependable tool to avoid other vessels in poor visibility.

The sun on the Strait hit a critical temperature, and the fog shot up from the water's surface, instantly reducing visibility to a few dozen feet. In less than a minute, my workload at the helm increased exponentially. To make matters worse, the current started to pick up. *Sea Hunt* was barely making steerage, and the radar and chart plotter displays were already jumping erratically with each shift of heading in the current. The unstable waters would abruptly catch the bow and spin *Sea Hunt* sideways, leaving an aggressive advance of the throttles as the only way to regain control. The moment the engines revved up, the towfish started bouncing at the end of its tether, making Ben's sonar image unusable. It was going to be a hectic, nerve-wracking day.

It struck me that we were likely repeating the Navy's most serious mistake. By not coordinating the survey of the trickiest areas around the slack current, our chances of getting a complete and accurate picture of the bottom were significantly diminished. The fog's obscuration of the horizon eliminated the reliable reference of peripheral vision, making the simplest change in course a chore. Reluctant to give up, we fought the ebb and flow of the current for nearly ten hours.

Finally, exhausted and with nerves frayed, I almost dragged the towfish into the reef. I told Ben and Rob that we needed to call it a day. Ben appeared reluctant but didn't argue. It was an awful feeling, knowing that I was the weak link in the operation. Even worse, I took away the exact wrong lesson from the experience by vowing to try harder the next day.

The current's real strength became evident only after the towfish was winched in and I was free to advance the throttles. Referencing the oil filled "wet" compass mounted flush in the dash, I pointed the boat due south, into the current and away from the shallows of Lawson Reef. The chart plotter display froze as the bow bobbed from side to side in widening arcs. It wasn't until *Sea Hunt* had accelerated to eight knots that it broke the harmonic-like swing of the bow, allowing the chart plotter's GPS receiver to re-cage on the boat's correct heading. We had been attempting to steer a precise course while barely making headway in just fifty-foot visibility. To do so without reliable chart plotter navigation data—while in the shipping lanes—was downright foolhardy.

Despite clear evidence of the impossible conditions, I started to second guess myself, rashly concluding that the out of control runs had been almost entirely my fault. Attempts to explain my erratic boat handling abilities—even to myself—were becoming unintelligible. Filled with self-doubt, all I wanted to do was get off the damn boat and breathe. Given the conditions, we had searched a tremendous amount of the sea floor; about ten square miles to date. But I was missing the big picture. I treated each hour with Ben and his sonar gear as potentially the last. It had felt as though we were constantly skirting the edge of disaster, and we would do it all again the next day.

I secretly wished that someone else would call, "enough," while at the same time ignoring the fact that I would never accept such a judgment. I knew I didn't look good, not with permanent bags under dully glazed eyes, not with my painfully gaunt features writhing like a dying spider. Back down to thirty pounds below my normal weight, with my body starved for calories and burning muscle, I was spiraling out of control.

What concerned me most was the prospect of a faltering confidence. Was I a mere pretender, ready to fold at the first hint of hardship? Or would I see the project through to the end, could I be a contender? It was a question I'd asked myself during difficult times before.

Master Gunnery Sergeant Bearup, United States Marine Corps, introduced me to the concept of "pretenders" and "contenders" in 1985 while attending Aviation Officer Candidate School (AOCS) in Pensacola, Florida. He had been assigned as AOCS's head Drill Instructor just a week before our graduation and commissioning as Ensigns in the United States Navy.

AOCS was a physically and mentally intense 14-week training program that was, in the final analysis, all "pretend." The academics, the brutal exercise sessions, the calculated abuse, were all conducted in a carefully controlled environment to build stress to near real-world levels to teach us how to survive as Naval Aviators.

There was one essential component of the training that enabled AOCS to rise above its contrived roots of pretense, challenging us to be contenders: the drill instructors. My drill instructor, like all the tightly screened, Marine Corps noncommissioned officers selected for AOCS duty, lived in a world of locked-up discipline that verged on religion. In fourteen weeks, my assessment of Staff Sergeant Gerhardt went from brutal sadist to mentor and hero, even though his actions were remarkably consistent throughout the experience. I was the one who changed.

Two images from the experience stayed fresh in my mind, each associated with a lesson learned on my final day at AOCS that I was only now, in the middle of the search for 510, coming to appreciate. The first was that of Staff Sergeant Gerhardt's enigmatic expression as I hand him the traditional silver dollar to convey my thanks for his training. His eyes are as serious as death, there's an almost undetectable curl at the edge of the lip, perhaps representing a grudging nod to life's penchant for irony; an aura of absolute respect absent the tiniest hint of the disdainful snarl offered by the same man just a week earlier.

The image moves. It is alive, and my drill instructor is ageless. Staff Sergeant Gerhardt exercises reality's cautious deliberation with a single hand, raised in the perfect edge of a salute. His fingers quiver in muscular tension as the salute reaches its apex and, with the barest acknowledgment of descent, the hand disappears as if a magician's trick.

It took me decades to fully understand that Staff Sergeant Gerhardt was not saluting me that day 29 years earlier. Certainly, he was paying respect to the new rank, but there was more to it—he was acknowledging my personal transformation. He was honoring my perseverance in making it through AOCS, saluting the person that I now knew I could be. Instead of paying tribute to a one-dimensional snapshot of fleeting achievement, Staff Sergeant Gerhardt was saluting the human potential that resides in all of us as we rise to challenge.

It wasn't until recently that I broadened my understanding of the second image to its proper importance. The scene was from within the AOCS barracks. With no fanfare or witnesses, Master Gunnery Sergeant Bearup calls our class to form up in the hallway. Master Gunnery Sergeant Bearup has been here before. The last time he walked these halls, it was with the same sense of noble urgency. Then, he faced the horrors of Vietnam. Many of the young men that took a similar printed card from him during that tour did not return.

He walks up to us slowly, hands us each a small card, and looking me square in the eye asks without inflection, "Which will it be?" I remember looking down at that card for the first of what would be many times in my life, in reflection if not always in reality. It read: "In life, there are Pretenders, and there are Contenders. The question is—which are you?"

The truth was that I'd been both pretender and contender at different points in my life, but as Staff Sergeant Gerhardt's salute taught me, what one has done in the past is not important; what matters is how you resolve to live every day of your life, starting now. My world had become a continuous sensation of running in deep, shifting sand, of fighting a losing battle with a desperately urgent outcome in the balance. Unimaginably powerful ennui had become my daily companion, a listlessness that was almost impossible to shake.

Fighting constant movements, painfully deep muscle contortions, debilitating fatigue, and anxiety attacks of dizzying intensity, I still marveled at how lucky I was, because unlike most people I knew (I was fairly certain), I was honestly and soulfully happy. As I took on each new challenge, I realized a fundamental truth carelessly tossed aside as a child: winning or losing really doesn't matter, so long as you did your level best.

Parkinson's disease is a sneaky son of a bitch, but I had endured, persevered, relentlessly refused to quit until the small victories were stacked high all around me, even though I was the only one who could see them. And then, the disease's churlish specter found a chink in my armor with a cleverly devastating surprise attack. I knew that my symptoms had grown to the point of distraction, but it was not clear at first what that really meant. I had been placing a GoPro video camera at different spots on the boat to help Dan Warter with footage for a future documentary. I watched the videos.

It was immediately obvious that my symptoms no longer affected just me; they directly impacted those around me. My nervous shuffle put all on edge; my gloomy struggle back from dystonia brought everyone down. I was a visual train wreck of distraction. And to top it off, my voice had become so soft and muffled that it was virtually impossible to understand me at times. I saw the video, and it was painfully apparent that these were legitimate complaints—it was damn hard to be around me. Parkinson's had found a way in. The disease was attacking a vulnerability: it was attempting to isolate me, to push away the relationships that make life what it is. And it looked like it might be winning.

For the first time since diagnosed, I was scared but also emboldened. I had learned tricks to get around past attacks, and I would learn new ones to beat Parkinson's latest end run. Twenty-nine years after AOCS, I finally had my answer—I was a contender.

Fourteen

...FOR THE FIGHTING MAD CORE TO PREVAIL

August 2014

It was shocking to think how close we'd come to overlooking the most promising sonar contact to date. Our best chance of finding the missing A-6 almost slipped from our grasp in the literal fog of on-deck excitement and danger. When Ben Griner, Rod DuFour, and I motored into the marina after *Sea Hunt's* near collision in the shipping lanes, the sonar return from our last side-scan run gradually receded in importance. It did not look like a jet. There were no visible wings or other readily distinguishable aviation attributes. But the contact was easily the largest structure painted by the side-scan within several square miles. The contact was also in the heart of the JAG Investigation derived high probability zone.

Dragging the towfish on the sea floor was more than a project setback; it was physically traumatic to me. As far as I was concerned, Ben was doing me a huge favor and I owed him. To be responsible for the damage or destruction of the deep-water side-scan sonar was one hell of a backhanded thank you. But it was more than that. I had come to believe that, given unlimited time, I could accomplish just about anything despite Parkinson's, maybe even because of the disease. My loss of situational awareness in the fog shook that supposition to its foundations; backing myself into that navigation corner rattled my confidence and self-image

Peter Hunt (at the helm), Rod DuFour, and Ben Griner during August bottom survey (from the author's collection).

to the core. In the widening pendulum swings of my distorted reality, the event made me doubt my core competencies—had a lifetime of experiences been for nothing?

Ben never once complained about the project's long hours or his personal financial investment in the project. During my daily fights with the disease, he held me to a remarkably consistent standard, in a way almost goading me to push back harder than I otherwise might have done. Ben acted the role of the "tough" coach who taunted, dared, and ultimately refused to give up on the struggling third string player, in the process elevating my performance to the point where the team had a fighting chance.

But now, feeling hollow, with no grounding or perspective, I struggled unsuccessfully to keep my head up in my lonely shuffle down the dock. Given the long work days, it was understandable that *Sea Hunt's* crew would usually split up quickly after reaching shore. A dread of being left alone at the marina lingers within me to this day like a cancer. Isolated, barely able to speak, helplessly falling down a steepening path toward futility, there was nowhere to turn. I became convinced—maybe correctly, maybe not—that most friends and acquaintances just couldn't handle my presence for long. The prospect of being left alone with my scrambled interpretation of reality on the day I dragged the towfish on the bottom was especially terrifying.

After a few minutes of awkward conversation at the edge of the marina parking lot, Rod walked across the street to his nearby home. Not knowing what else to do, I went into the Marina store to buy a couple of beers and pastries—we had not eaten in the last twelve hours. I offered one of each to Ben, fully expecting him to decline and continue walking to his truck. To my surprise, he took both, sat at the picnic table in front of the store, and talked me down. He just sat there, listening, occasionally commenting, with no regard for the long drive ahead of him before reaching his home and family. Through a simple act of kindness, Ben stopped my plummet into the darkened pit of despair. I brushed myself off and got up.

During the few days it took for Ben to refine the sonar picture, my excitement slowly grew. Scrutinizing the sonar return, I looked for the smallest clue to prove to me that the long, grueling days on the water had been worth it. What the hell was this thing? At first, it didn't look like anything recognizable, but after staring at the image, some thoughts began to gel. The general condition of the contact seemed to fit a logical pattern. It would have been hard to find in the dead of winter, located as it was in a zone of extreme current, open to winds and storms from every direction. It was also a mile and a half away from air traffic control's projected crash site. If the Navy had prioritized their search based on the JAG Investigation's radar data conclusions, which was counter to what the *Salvor* report indicated, but was still possible, then this new contact could have easily been overlooked.

What sections of the Intruder might still be relatively intact after the crash and 25 years of immersion in salt water? I stopped hoping for a hit-the-jackpot identification point, like clearly visible wings. I reviewed the aircrew statements again, trying to visualize what 510 would look like today.

Five-ten was configured abnormally when it hit the water. Therefore, it shouldn't look exactly like an A-6 after a normal landing, even if it were completely intact. Five-ten's main landing gear had been stuck up with doors closed flush with the fuselage. The nose landing gear, however, was at least partially down. If 510 was sitting upright, then the jet's nose might be marginally higher than the remainder of the fuselage resting on the bottom. Any part of the aircraft that was taller would display more shadow on the side-scan sonar image.

I forced my eyes away from the computer monitor, focused on an empty spot on the wall, waited a few seconds, and then looked back at the sonar image. Something was there. At the top of the sonar picture was the shadow of a faint, but well-defined, line with a 90-degree bend. Nature doesn't typically place objects with perfect right angles underwater, certainly not in the middle of a vast barren patch of sand and small rocks. I closed my eyes for a slow three count and opened them. I felt butterflies, not dyskinesia, but an honest excitement in the pit of my stomach. Could that possibly be 510's aerial refueling probe, the most defining characteristic of the ugly old bomber? I had been working under the assumption that the refueling probe must have broken off on impact. Could I be wrong?

Wracking my memory for a reference point, I tried to think of an example of the probe's strength. The fiberglass aerial refueling probe extended in front of the cockpit, almost directly at the point where water impact occurred. It would have been exposed to the greatest force of the crash. Logic said no, there was no way the probe could have survived the crash. Still, it felt as if I was missing some key data point about the refueling probe.

The answer came in a flash with a visualization of the back of a giant Air Force tanker, a KC-135, an airliner-sized variant of the Boeing 707. I remembered all the times hundreds or thousands of miles out at sea training for long-range strikes with the four jet-engine tankers. The Air Force conducted tanking procedures differently than the Navy. Air Force tankers had a rigid boom that stuck down and out the underside of the tanker aircraft at a 45-degree angle to the rear. An observation bubble with a crew member inside was located where the refueling boom connected with the tanker. The crew member's job was to steer the boom into a refueling receptacle at the top of all Air Force fighters.

The Navy operated with the opposite philosophy, and the Air Force had a barely acceptable modification to allow the transfer of fuel to Navy jets. Before taking off, the KC-135's refueling boom would be rigged to accommodate a ten-foot-long hose terminating in a rugged, one-hundred-pound steel ring. Navy carrier pilots called the KC-135s "probe-busters." Navy pilots needed to fly their aircraft's probe into the center of the unforgiving steel ring, push it in far enough to form a small bend in the short refueling hose, then stay locked in position until the desired fuel was transferred. The two aircraft would fly with less than three feet of play between them, through clouds, while turning and in turbulence. It was not an easy maneuver, and it was not uncommon for a jet's probe to be broken off due to inadvertent lateral pressure. But just as often, the Intruder's refueling probe would win the battle, tearing the sturdy basket violently from the tanker's boom. The A-6 refueling probe was strong enough to break the KC-135's steel ring free, fly with a length of hose and its steel rim flapping in the hundreds of knots of relative wind back to the aircraft carrier, and then decelerate from 120 knots to a standstill in a second and a half when trapping aboard. Crashing into the ocean at 150 knots probably wasn't any more stressful.

I identified points on the sonar image for Ben to measure. If the shadowed line was 510's refueling probe, it should be about three feet long, make a 90-degree turn, and continue on for another foot and a half. There was also a shadow at the top center of the contact. Could it be the holes in the canopy where the crew ejected? The remainder of the image looked smooth, as it should with 510's landing gear stuck up and the main gear doors flush with the fuselage. The fuselage itself should be about ten feet in diameter at its widest. The length of the fuselage could not be any longer than 55 feet, but depending on how it might have broken at water impact, it could be shorter.

Excited by the possibility, I shot an email to Ben and Dan Warter explaining my thinking and requesting the various measurements. Ben replied the next day.

Friday, August 22, 2014

Guys,

Here are the measurements. There does not appear to be anything in the sand around the target. There is a clean edge on the back side. The tail would need to be missing as there is not a shadow. The hole in the top is 2.5 feet by 2.5 feet, but it's hard to measure, so let's say 2 to 3 feet at the edges. The end opposite the hole seems to have some significant and deliberate angling that would be odd for a marine vessel. I have been wondering if we are looking at the front part of the fuselage. How large is the front landing gear door? Could it be open?

Ben

Even after Ben's encouraging email, the absence of visible wings bothered me. It was possible they tore off at water impact. But even if they hadn't, with the fuselage flush to the bottom, as was probably the case, could the wings be buried in the sand? For months, we had marveled at the ripples along the sandy bottom formed by the swift current. It made sense that the wings had been covered by sand over the years.

The next day I met with Tugg to get his take on the sonar image, and he quickly provided an alternate theory. Might the shadow with the 90-degree bend be the partially extended nose landing gear? Could the hole that I interpreted to be a break in the canopy glass instead be the open nose landing gear doors, as Ben had suggested in his email? If so, the wings would have had to shear off in the crash, leaving 510's fuselage lying on its back. I grew excited: two reasonably plausible theories in two days, both identifying the contact as the missing A-6.

After reviewing the contact's dimensions, Ben shared his confidence that the sonar return was indeed 510. But we couldn't know for certain until we had eyes on the target. This presented a major problem: the tides were too strong to make an identification dive for at least another two months, which started to encroach upon my DBS surgery date.

Fortunately, Dan Warter had an alternative. He knew a professional ROV owner/operator who might be willing to video the contact. Craig Thorngren graciously offered the use of his ROV's services, but our request would have to fit his tight schedule. He could dedicate a single day of ROV operations to the project before leaving for a job in South Africa.

Craig is a retired Coast Guard Chief Petty Officer who spent his active duty primarily working law enforcement on the high seas. He was used to operating ROVs from the decks of large ships, and *Sea Hunt's* limited space must have thrown him. The massive winch on the cramped aft deck probably didn't help make the transition any easier.

The inflexibility in ROV scheduling presented an additional problem. The sonar contact was smack in the middle of Rosario Strait's western set of shipping lanes. We needed to coordinate with the Vessel Traffic Service and receive permission before Craig could deploy the ROV. The standard request process for ROV or dive operations took a minimum of ten days for approval, and we had just two. My first emailed request was denied. Over the next 24 hours, I submitted a second request to the VTS, composed several explanatory emails, until finally, as a last resort, I called the Director of the Seattle VTS sector personally to try and convince him to change his mind.

The phone call was particularly harrowing, as it was late in the day and I wasn't at all confident that my voice would be intelligible. Eventually, after twenty minutes and on the verge of pleading, I explained to the VTS Director about my disease and the urgent nature of our request.

"This might be our last shot at identifying the contact," I told him bluntly. "It's an all-volunteer crew, and we're all losing patience."

The VTS Director, apparently sympathetic to the situation, went to great lengths to integrate the ROV search into the shipping schedule for Wednesday, August 27. This left me the early morning to find and install several pieces of specialty equipment, such as a radar reflector, needed to satisfy VTS requirements to operate in the shipping lanes. Arriving at *Sea Hunt* moments before Ben, Dan, and Craig drove into the marina parking lot, I started the day—as usual—exhausted.

The ROV could operate in a maximum current of three knots, but that assumed a fixed staging platform, and the VTS agreement restricted us from anchoring. This was to ensure the ability to maneuver quickly if an unresponsive ship approached the site. *Sea Hunt* would be hard pressed to get on station in time for the brief slack tide, but none of the others seemed concerned. Despite the calm and clear conditions, even the smallest breeze or current could quickly push the relatively light-weight boat out of position.

Once on site, we made pass after pass, attempting to drift by the contact slowly enough to get the ROV in place for video imaging. Dan Warter would furiously let out the ROV communication tether by hand as the mini-sub raced for the bottom. Meanwhile, Craig pinged the surrounding waters with the ROV's onboard sonar from the laptop in *Sea Hunt's* cabin, furiously searching for the contact. After just three minutes in the current, the ROV and *Sea Hunt* would be pushed hopelessly out of position. Dan would then pull in the hundreds of feet of tether and the ROV to try again. After an hour and a half of attempts, we checked out with Seattle Traffic and cleared the area. Our best shot at positively identifying the sonar return in 2014 was gone.

Sea Hunt's aft deck during the ROV identification attempt. Dan Warter, Craig Thorngren, and Ben Griner hauling in the ROV (from the author's collection).

The Lost Intruder

At a temporary impasse, but wanting to take advantage of the borrowed winch, we decided to look for a shipwreck that had long been on Ben's "to do" list. We were joined the following day by deckhand Paul Hangartner, an MDS diver I had met briefly earlier in the year. Paul is a tall, lumbering aircraft mechanic with a shaved head and shaggy goatee. Generally good natured, Paul had an edge to him, a kind of tough-guy compulsion to put people on the spot with uncomfortable conversations.

Paul had been an enlisted man in the Navy, and he started the day with a long rant on the privileged nature and general incompetence of officers that was undoubtedly aimed directly at my past affiliation. He waited to gauge my reaction, but I gave none, and instead did my best to remain affable. The deep diving crowd has always been on the rough and tumble side, and it was common to give other members of this exclusive cadre a hard time. But Paul's critique seemed different: it was more thoughtful, more serious at its core. There was something eminently likable about Paul, and I didn't take any of the critiques personally. After knowing Paul for a while, it gradually dawned on me that there was a method to his madness. Paul chose the topics of his pointed observations carefully, each seemingly orchestrated to shake the foundation of assumptions on which most of us base our lives. From the wars in Afghanistan and Iraq to politics to daily rituals such as going to work, he worked diligently to tear down the self-constructed illusions of individual identity.

By highlighting what were often ill-considered priorities in our lives, Paul demanded a degree of honest self-appraisal, often shining light on the real motivations for our actions. Paul was a truth-teller, an outlaw of ideas unabashed in his analysis of the false gods we happily erect as private prisons. I learned to appreciate Paul's outspokenness and saw it as a refreshing appraisal of unexamined lives.

He was also loyal to a fault. As Rob Wilson put it later: "You can trust Paul with your life at 300 feet, just not to get a pizza." We became friends.

There was more to Paul Hangartner's story. In an amazing coincidence, Paul had served in Attack Squadron 145 as an airframes mechanic. He left the Navy in 1988 shortly before I joined the squadron. Airframe mechanics were responsible for maintaining the A-6's hydraulic systems and, in fact, Paul had worked on 510. An avid deep wreck diver, Paul's involvement in the project was another long-shot coincidence that didn't strike me at the time as being the least bit strange.

How fitting, to have Paul as one of the first divers to touch the lost Intruder. There was even an outside chance that he might discover the reason for 510's hydraulic failure, or perhaps why the main landing gear did not extend. Admittedly extremely unlikely given the decades of aircraft decay, it was still in the realm of possibility. There had to be few, if any, other technical divers in the world who had been trained to work on the A-6 hydraulic systems. And there was exactly one technical diver on the planet who had actually put a wrench to aircraft 510. To have him here, on the team, was nothing short of incredible.

Sea Hunt cruised for the two and a half hours south to Port Townsend on the Olympic Peninsula to look for Ben's wreck. Despite the clear weather and calm seas, conditions were still challenging. The Whidbey Island to Port Townsend car ferry ran every hour directly through the sector we were searching, and swift currents pulled unpredictably at the shallow coastline. Mid-way through the survey, shaking uncontrollably and with a horrible anxiety screaming through my brain, I couldn't go on. I abandoned the helm to Ben. For the next two hours, Paul operated the winch while Ben ran the sonar and drove the boat. I sat idle, frustrated, angry at myself; ashamed at my deteriorating abilities. Finally, we found the wreck, marked its position and it was over. It was the day that I finally fell apart.

It was not the way I envisioned ending the summer. It was beyond embarrassing, even worse than dragging the towfish on the bottom. With my resilience shattered, feeling utterly ground down, I finally recognized part of what was happening to me psychologically. It took the head-on impact of a breakdown to prove to me that, just as with every healthy person, my limits might be hard to discern but they did exist. My relinquishing of the helm, although the correct decision, was also one of the most difficult in my life. It is an exceedingly thin line, but pushing back blindly at a disease can be just as harmful as going gently into the night. It is a lesson I try to view positively and do my best to remember every day. It has proved to be a particularly sharp double-edged sword.

The Lost Intruder

I took back control of the helm for the return run to the marina. We pulled alongside the bulkhead at the 8:00 pm high tide, just in time to meet Captain John's crane, ready to finally remove the cumbersome winch from *Sea Hunt's* deck. As the sun went down, I stood on the empty aft deck, wiping up hydraulic fluid residue with a mop of oil absorbent pads underfoot, watching the crane get loaded into the bed of Ben's truck. Once Ben and Paul had left the marina for home, I took the boat back to the slip by myself. I backed it in and tied the bare minimum of lines before collapsing on the settee in exhaustion.

With Craig Thorngren out of the country indefinitely, and probably reluctant to work from such a small boat again, getting divers on the site took on a renewed importance. The days were shortening as the Pacific Northwest autumn approached. It seemed unlikely that we would be able to find a date with satisfactory currents and enough daylight hours to span the required decompression of a 248-foot dive. But I also knew that it would be damn difficult convincing the MDS divers to risk their lives from my boat when I couldn't even be relied upon to tow a cable. They were right, I couldn't even convince myself anymore, and two weeks later I stopped trying.

This sort of unfinished business used to bother me, but with DBS surgery nearing, my thoughts turned to other matters. Would the procedure change me as a person, and if so, would it be for the better? So far, the disease's progression had been linear, but that bedrock of continuity might be about to disappear.

Would I forget the hard-fought lessons of the last year? It was probable that at some point in the future my condition would deteriorate again. Would I remember the strategies to deal with the Parkinson's symptoms learned on the first go around, and if so, was this even desirable? Would I still be "flying the plane," or was DBS a deal with the devil to scratch more time out of a life of questionable quality?

Nearly everyone judged my happiness through my physical appearance. When obviously symptomatic, I was assumed by others to be leading a miserable existence, my protestations of happiness glumly ignored. It was infuriatingly frustrating. Was anyone listening to me? Was I only now, 52 years into life's journey, noticing this? I honestly didn't know.

The prep for surgery started with a pair of MRIs at Swedish Hospital in Seattle the month before the procedure. Just three weeks out from surgery, my afternoon and evening energy level spiked. I worked out at a ferocious pace. Jared drove me to a pair of school board meetings, where I successfully fought to stay engaged. With my handwriting now illegible, even to me, I painstakingly keyboarded memorized notes in a clipped shorthand after returning home.

I tried not to dwell on the possible surgical outcomes, finding solace instead in the lost Intruder project. It was reassuring to know that the missing A-6 would still be a mystery when I emerged from brain surgery. It was comforting that at least one concrete purpose for existing these last eight months would still be there.

There was one final pre-surgery challenge to overcome—hyponatremia. Lab reports indicated a dangerously low sodium level from drinking excessive water due to perpetual dry mouth. If I could not control my water intake, the surgery would be postponed. I lived in a constant thirst for the next two weeks. The sodium level blood results came back barely satisfactory as I was being wheeled on a gurney into surgery.

I said goodbye to Laurie and Emily in the surgical prep area. Jared had a school commitment, and he shared his father's understanding of the importance of keeping regular those things in one's control. I smiled awkwardly as I chewed an ice chip and the anesthesiologist had me count backward from ten. I made it to eight before sliding into darkness.

Fifteen

Sudden end to intensity's shudder…

Autumn 2014

Parkinson's progression is like "boiling a frog": life's natural coping mechanism, time, tends to mask the real weight of the illness from the afflicted. But imagine if the emotional distortion of time did not exist. What if one could instantaneously transit the full range of Parkinson's most debilitating symptoms accumulated over a decade, and then with the literal push of a button set back the clock ten or more years with almost no indication of the disease? This is where I find myself at Christmas 2014, recovering from Deep Brain Stimulation surgery; confused as hell.

I am now physically different and will continue to be, presumably, for the rest of my life. Without thinking, my hand gently runs over the top of my head until I feel the set of ridges left by the procedure, the pair of two-inch breaks in the skull that are almost healed less than two months after the first incision. It never occurred to me that the incursion into my brain, where the two separate wires—one into the left side of the brain, one into the right—were inserted would be the less painful, far easier of the two surgeries in recovery. I can readily imagine the dual bald spots, highlighted by the recent growth of hair that surrounds them like the top of a teeming anthill, where the ground goes bare as the lines of foraging ants disappear into the colony's mound.

The scars are also bare on the inside. Not really, as they each expose an electrical wire, but I've come to think of them like that. It's my way of understanding how they work, my visualization of the electrical signals sent through the wires to disrupt the random pulses sent out to the body without the regulating moderation of Dopamine. It makes a contorted logic. Bad signals from deep within the brain produce uncontrollable outcomes. The left side of the brain affects the right side of my body and vice versa. It works well enough, perhaps too well, I think. But again, how will I know for certain? Does it even matter?

After the second operation to implant the batteries in my chest and connect the wires routed under the skin in my neck, the surgeon left me with a six-inch-wide racing stripe of a week's worth of new growth hair at the top of my head. It looks like a Mohawk. I wonder if he did this on purpose, secretly hoping that he did. This would signal that my sense of humor was appreciated during the surgery. It harkens back to an old Naval Aviation saying—it's better to die than to look bad.

My eyes scan to my left hand as I watch it rise to join the right in a miraculous coordination of movement. The steady hands come together on a parallel plane in front of my chest as I marvel at the question, how does it know? How does each wire identify which brain signals are wrong, which ones need disruption? I relegate that mystery to the bin of miracles witnessed at each "tune up," my three visits to the Advanced Registered Nurse Practitioner, "Heather," I'll call her. She adjusts the voltage and other electrical parameters until the system is most effective in reducing specific symptoms.

My hands go to the top of my chest, one to a side, barely brushing the skin on their downward travel to the raw, still healing dual scars, the swollen bulges of the pulse-generator batteries just beneath the skin. Ten months later this will develop into a habit whenever doing anything physically demanding. One year later, I will have almost grown used to them; but not quite. But it will take years, I suspect, to shed the "deer in the headlights" feeling of self-conscious awe. My hands trace the wires beneath the skin to where they each join their own two-inch battery, comparing the entry points, trying to determine if either has broken free. I do this somewhat paradoxically without thinking.

The Lost Intruder

These are the power sources for the wires that have been inserted expertly around parts of the brain, without ever penetrating the gray matter, until they are secured deep at their destination. I find myself thinking a lot about how the pulse generators will need to be replaced in about four years. I wonder if all my working out will be for naught once again due to the post-surgery recovery, downtime that will weaken my body and will. It's a ridiculous thought, I know that, but it still haunts me. My future dependence on replacement batteries, the least invasive part of the procedure, troubles me more than it should. As if life wouldn't be incredible with possibility were this my only valid concern for the next four years. Two weeks later such worries disappear as I unwind the core unease of my twisted disquiet. Must I eventually go through the full descent into Parkinson's hell again? Of course I will, recognizing this unasked question for the first time.

The first time through was a new challenge. How will I respond to an already experienced hardship? Seven weeks after surgery and already I find the need to haul my focus back to the present, where thoughts have influence; where they belong. Perhaps the most valuable lesson learned in the past few years, and threatening to disappear in a matter of weeks, is to live in the present. There is nothing else. Suddenly, I am nervous, but now with clear reason. It is a night and day change from the randomly free-floating Parkinson's anxiety.

The pulse generators are prominently visible when shirtless. They look like alien implants, have the firmness of their titanium steel, and the imagined warmth of a hard-working battery. The wires each run along the side of my neck and under the scalp to the still tender folds at the sides of my head. These two modestly painful reminders of surgery are covered by scalp and hair. The bulges made by the wires are clearly visible but easily mistakable for veins up to the hairline. The pulse generators feel warm and itchy and will continue to for months.

I can't feel the electricity releasing in my brain, but the pulse generators are always a concern. The batteries are the system's potential weak point; they are where a break in the wire would most probably occur. A softball line-drive or the recoil from a rifle, Heather relates to me from experience with other patients, can disengage the wire. But it does take a lot of force. I can practice hot yoga, lift weights (no bench press, however), and maybe start running again. The list of restrictions is short—no hot tubs greater than 100-degrees Fahrenheit and no scuba diving below 33 feet. I don't consider the diving limitation a hard and fast rule. It's still my body, and I'll do with it what I please. But exceeding 33 feet could have warranty implications, which makes me think of all sorts of unrelated and irrelevant medical ethics issues. MRIs and a pair of less common medical procedures are also prohibited—it's all on a medical alert dog tag around my neck, which I can reference if I really need to know. This lasts just three months before the necklace is relegated to a dresser drawer forever. I've never been much of a jewelry guy.

I make a mental note to call the medical device technical representative after the holidays to learn more about the 33-foot diving limitation. It sounds like an arbitrary limit, as it corresponds to exactly one additional atmosphere of pressure. Heather tells me of one other diver with the implants who has gone to 80 feet with no ill effect. Still, if I decide to venture beyond 33 feet, which of course I will, it needs to be done incrementally. But I know this "walk before I run" philosophy will vanish with one glimpse at the edge of the visibility of an unknown shape or fish—or who knows what—that will drag my curiosity further down. It is immensely reassuring to know that in at least one major way, I have not changed.

The brain surgery was not a pleasant experience. It was far worse than I expected, being completely out of control, with my head clamped securely in place by a cage and what felt like a dozen pointed screws fastened tightly to my head. The anesthesiologist put me out just before the catheter insertion. When I awoke, it was to the incongruous buzz of an electric razor and the comment, "That's the third broken razor because of this guy's hair!" To my surprise, it was the surgeon who was doing the shaving. I mumbled into the operating room's twilight darkness—my sedated interpretation of the lighting, at any rate—the question, "Why is the highest paid guy in the room doing the shaving?" Either nobody understood me or chose to answer. Or laughed, for that matter.

The Lost Intruder

I suspect my memory of the surgery will last forever, most notably the minute or so of breaking into the skull by what felt like a jackhammer. "Try opening your mouth," Heather reminded me, holding my hand to get my attention. "It might dull the vibration." Dutifully, I opened my mouth wide. It makes no difference, but it does distract for a moment; it gives a fleeting illusion of control.

The memories of the nearly three hours of wakefulness during the brain surgery would haunt if allowed. Each identical procedure was conducted by a different surgeon: the left side by the novice as a part of his on-the-job training; the right by the 17-year veteran of DBS surgeries. I watched the first half—the left side of the brain—on an overhead video screen, not wanting to miss this once in a lifetime experience. But, I start to think, as the scalp is slowly peeled back at the top of my head with an audible "rip," that maybe this isn't such a red-hot idea. A sharp, searing pain, immediately followed by a throbbing, dull ache is choreographed perfectly to my scalping. I cry out, "Doc, more Lidocaine please!" I intended to say more, to ask more questions, but quickly realize that I am way over my head, literally, and maybe watching the procedures is a mistake. The left side surgery seems to take forever. Heather makes small talk to try and keep my mind off the penetration of my skull.

With my brain now exposed, the video of the surgery becomes clearer, although maybe my eyes are blurry from anesthetic because while I can distinctly see the new break in the skull, the brain itself is not apparent. There is no sensation as the wire is fished around the various lobes and protrusions, while it is inserted deep into the gray matter until the bitter end reaches its destination. I wonder how they know: does the wire stop after meeting resistance, is it measured out ahead of time, or is it something else entirely that I haven't considered? The unsettling thought suddenly strikes me that there are quite a few things about the surgery that I had not considered.

"Watch the fingers." It is the experienced neurosurgeon. While not anxious or panicked, there is urgency in his voice. I imagine the novice's hands slipping. His fingers are no more than six inches above my center of thought, but probably due to the video projecting on the ceiling, my impression is that the surgeon's fingers are a dozen feet away. The cage that locks my head in position feels huge as if it is a giant cube with four-foot sides. I can't see any of the cage, but the sensation is that it surrounds my face, a definite challenge for the claustrophobic.

The skull incision is irrigated by a nurse with a squeeze bottle. I hear "Catch that!" as a warm fluid runs down the front of my head toward my eyes and mouth. Hands reach between spokes in the cage, trying to dab and dam the flow before it enters my mouth. Do they think tasting a bit of the fluid—presumably synovial, I remember from some long-ago anatomy class—will bother me after feeling my skull broken open? Some does reach my mouth, and there is no taste. This is the one surgical incident that doesn't bother me at all.

With the first wire placed deeply in my left brain, Heather releases my hand and picks up a small boxed electronic instrument. She adjusts the settings and then touches the wire going into my head with a lead from the box. A loud "wuushhh," followed by soft, muffled noises, like the dulled sounds of whales communicating underwater, fills the operating theater from some unseen speaker. Heather mentions that the signals are firmly in the normal range; I successfully fight the urge to make a smart-ass crack. Barely.

The anesthesiologist is in my line of sight, sitting about ten feet from the foot of the operating table, far enough away that his mask is below his mouth. He looks bored. There is only one interaction between the anesthesiologist and the surgeon that I can remember, but the words don't register for some reason. He just sits, occasionally squirming uncomfortably.

More adjustments. My right side seems to relax in response, so I verbalize the sensation. This is the reason why the patient is awake during DBS surgery. It is only through anecdotal confirmation of the resolution of symptoms that the precisely correct wire placement can be judged. It seems to take a long time. I do my best to verbalize every sensation, but no one appears to be listening any longer. I think about stopping, but prefer to stay preoccupied as the head cage is becoming confining; a tinge of claustrophobic panic searches for a toe hold. Memories of being lost deep within the blackened wreck of the *Andrea Doria* predictably threaten. I focus on my symptoms. The electrical current in the wire is working at its crude setting, I can tell that it's reducing some symptoms, although it's difficult to pinpoint exactly which ones. I just feel better.

After the wire placement on the left side is complete, I ask Heather to turn off the video. One of the surgeries is enough to watch. I have the option of stopping the surgery after the left side is complete, but I say to continue despite badly wanting it to stop immediately. It is painful, extremely uncomfortable with my head immobilized; with the cage's screw points gouging into my scalp to keep my head still. These are the first wounds to heal, not those from the surgery itself, but rather the small cuts in the scalp where the screws were cinched down until they bled.

Immediately before the operation, I underwent a special CAT scan to check for "brain atrophy." This would have been a disqualifier for conducting both sides of the surgery in one procedure. Brain atrophy allows too much movement in the skull, making it far less sure that the wire leads are going to the proper spot. Fortunately, there is no brain atrophy. It is a dubious achievement of which I shall brag to unresponsive audiences for months.

The novice surgeon finishes closing the break in the skull by literally sprinkling pieces of bone back into the void, like a kid trying to cover up a mess. I'm told the pieces will somehow heal naturally into a smooth feeling ridge. And, sure enough, they eventually do. The right-side wire insertion goes much faster, as it is implanted by the experienced neurosurgeon. A representative of the medical device company is in the operating room observing. He becomes helpful towards the end when he and Heather pepper me with questions to take my mind off what is happening. Before I know it, the right side is complete. The screws loosen from the cage around my head. The next thing I know there are images of the passing hospital ceiling as I'm wheeled to the neurological intensive care unit. I am lifted from the gurney onto the bed where I will remain, forbidden from rising or walking, for 24 hours. Fortunately, the surgery leaves me mildly constipated. I have no idea how a bed pan is supposed to work for a bowel movement, and I'm damn glad that I never found out.

The rest of the day, it is only 11:00 am when I reach intensive care, and night is spent ordering food, eating, and dozing off. It is far from unpleasant. There is a dedicated Registered Nurse for every two patients in the ICU, food for my famished body delivered on call, and a comfortable bed with air cushioned splints on my legs to avoid soreness from inactivity. Reading, however, is problematic. Placing the stems from my reading glasses on the sensitive scalp behind my ears is exceedingly painful, so I break the frame between each lens to use as a set of monocles. I surf the internet on my smartphone for ten minutes, doze off for ten, briefly attack my tray of food, and repeat. The clock moves incredibly slowly. Everyone is over the top helpful, cheery, and I start to enjoy the stay, even feeling a bit guilty for being in the ICU. This is especially true after one man—who it is not difficult to imagine is at the edge of death from the sound of his moans—is wheeled into an adjacent room in the middle of the night.

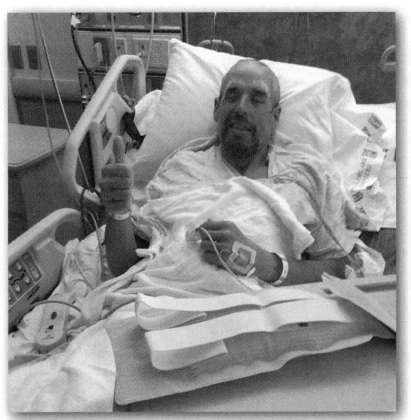

The author in the neurologic intensive care unit one hour after Deep Brain Stimulation surgery (from the author's collection).

The Lost Intruder

Immediately following the brain surgery, I experience the "honeymoon effect," a not uncommon occurrence after a DBS procedure. For the better part of a week, I do not show any Parkinson's symptoms whatsoever. But they slowly come back. Within weeks, I have re-lived the entirety of the disease's progression since diagnosis. It is a "Flowers for Algernon" period, physically disorienting and confusing.

Laurie and Emily come back to the hospital the morning after surgery. Check out is the only slightly rough patch of my stay, with the neurology ICU staff understandably inexperienced in checking patients out of the hospital. I conclude that they are likely more accustomed to processing those who have died than reasonably healthy folks. I decide to not even joke a complaint.

My last medical check is with a physical therapist who makes sure that I can walk. I've spent less than 24 hours in the neurology ICU. After being wheeled to the hospital sidewalk, I finally stand and move to the car under my own power. I sit in the front passenger seat with the shock of a newborn leaving the hospital, the focused attention of dozens of professionals deserting me under the influence of an unseen timetable. With the signing of my name, the confident nurses are replaced by a struggling family desperate to help. We muddle through.

The hours go slowly as the days fly by. In less than a week I'm back at Swedish for the second of the pair of surgeries to install the batteries and connect them to the wires to my brain. It is out-patient surgery with no overnight stay required, and I leave twelve hours after arriving. Steve Kelley, the friend who paid for half the Dragonfly sonar, drives me both ways. On the way home, I place one bag of ice on my chest and another on my head to where still more scalping has occurred. This procedure adds two more ridges to the sides of my head to make room for the routing of the electrical wires that then travel beneath the skin of my neck. Time is destabilized, not acting predictably, or am I just now noticing the fragmented progress of time that has always surrounded me?

Three weeks go by since the wires were implanted in my brain. Three weeks of feeling lethargic, beat up, of forgetting what feeling right or feeling "off" ever meant. I wake up every couple of hours with "REM-Wood," a juvenile description I've coined for waking up with an erection, something indicative of being in "REM" sleep, as I've learned from a friend in a decidedly un-lofty conversation. I view REM-wood enthusiastically as a positive sign. It is one of the few things worthy of excitement for weeks to come.

Friends and family ask if I'm nervous about the battery turn on date. I don't know how to respond. There are worse things than dying or being immobilized physically. Having your spirit stand still, not being able to try anything, not being allowed to push back, is the worst of all. I've got to be allowed to play the game. Otherwise, there is no point. I realize, of course, that this is probably just temporary, but it reeks with all the oily brimstone of hell.

I seek old solutions to familiar but reformatted problems. I look to *Sea Hunt* in search of freedom and choose a challenge of necessity, changing the boat's fuel filters. It goes well, which seems to get me back on track, at least temporarily. The dock water is turned on again after the early freeze of two weeks ago, so I hose down the deck and the window rails.

I take *Sea Hunt* out solo but wear my life vest in a nod to my weakened state. The water runs swiftly in Deception Pass. The boat controls well. It feels good. I know this is a place to return to for centering if needed, the familiar waters. If the DBS does not work, and in the worst case my abilities don't rebound, do I still have the drive to fight back? I am so damned tired. Accepting life as it is can bring happiness, I remind myself. But it sounds empty. Truths learned just weeks earlier now seem to fall flat.

It is finally the eve of the day the pulse generators are turned on, and my mood is relatively optimistic. Sleep has been difficult due to the ridged incisions on my head and the twin cuts on my chest, but once lying in bed, the mattress is truly comfortable for the first time in years. I thought at first that this was just a byproduct of the pain medications, but it's been almost three weeks off pain killers; four weeks since the cranial surgery. Each night I still marvel at the softness of the mattress. Lying in bed is the most enjoyable physical thing experienced in a long time, even more so than REM-wood. And my sleep reflects this. Even when just catching cat-naps due to post-surgical pain, it is deep, REM sleep, a restfulness that has eluded me for years. This aspect of the honeymoon effect stays with me until the battery turn-on date.

Laurie drives me to the hospital and, before I know it, I'm swinging slowly on my crutches, moving under my own power, as I had vowed, through the main entrance doors. Without medication for twelve hours, my right-side dystonia has returned completely. I need to clear my body of Levodopa to present a clean slate for Heather today when she powers up the pulse generator batteries. She comes out to the waiting area to bring Laurie and me back to the exam room. She asks about the crutches. I forget that she joined the team mid-way through my screening for DBS and has not seen me off meds before, except when being wheeled to and from surgeries. She accepts my inability to walk unaided as if it were an everyday occurrence, which for her, I suppose, it is.

Heather's enthusiasm is infectious, and it turns out to be the perfect segue way to the positive mood of battery power up. The pulse generators begin to operate, and the crutches are put aside. With the flip of a switch, my life changes: I can walk unaided; I can breathe freely.

And then it is time to turn the power off, just for a few minutes, to accomplish a calibration of the settings. It must be shocking to watch—the change is immediate and severe. My right-hand trembles uncontrollably, both legs kick out repeatedly, and I can't sit still. But the most frightening thing is my face. My eyes become clouded; my mind is fuzzy. I hunch over as though there is a hundred-pound weight on the back of my neck. It becomes difficult to breathe.

Heather apologizes when we must temporarily turn off the pulse generators. Invariably, I am confused but touched by the apology. This turns into my favorite part of future appointments. Instantly, I am transported back to the painful, heavy load of an existence developed over ten years. And just as quickly, in the time it takes for a full, deep breath, I am back to the "old" Peter again: sitting straight, eyes wide without pain, alert and alive. I never want to forget both feelings, even though I'd prefer to spend my life in the "power on" state. I am fortunate beyond words for so many things, not the least of which is the ever-present reminder of how difficult life can be.

Heather offers a celebratory hug, and I accept it gladly, feeling as though I'd shared the entirety of the DBS experience with her. Suddenly, I'm overwhelmed with emotion. I break poise for a moment as my eyes grow moist. I quickly regain control, but welcome the memory of honest emotion after so many years of clouded, toughened sensation.

At first, there is only the slightest hint of concern in the back of my mind. It almost goes unnoticed. But within days, the questions creep from the recesses of my consciousness, slowly emerging. The doubts start to nag, to drag me down, not with any distinct pre-surgery threat, but vaguely, like a bobbing and weaving prize fighter that I can't seem to lay a glove on.

Will those hard-won markers on the road to happiness be forgotten? Have I changed permanently, who am I, now? Where does the lost Intruder fit into all of this?

The last turns out to be the only readily answerable question. Now armed with the exact latitude and longitude coordinates of last August's contact from Ben, I can easily find the wreckage on the Dragonfly DownVision.

One week later I am on *Sea Hunt* again, solo, checking the reality of the tidal swing versus the published predictions overhead the contact. It is close, within ten to twenty minutes. In the coming months, I head out to the site half a dozen more times, wanting to ensure the groundwork is firmly in place for the MDS crew when the currents, tide, and the weather finally allow for an identification dive. I am positive that the contact is the missing A-6.

I'm not sure of anything else. If the fatigue just leaves, even for a few days, then I've got a fighting chance. At my core, I know that no matter what disability, no matter how tired, worn out, or beat up I may become, that it is—at the end of the day—my life. It's that simple. I know this intellectually and intuitively. But am I able to break the freeze of comfortable indecision and act? Or has it been a deal with the devil? Fuck it, I try to tell myself; there is no devil, at least none not contended with before.

I vow that if it has been a bargain with the devil, then I still owe my courageous pre-surgery counterpart of past battles the courtesy of acknowledgment. If I can do that, then who is to say what else I can control? I must fly the plane.

With that thought both prodding me forward and beating me down, I look to the familiar for inspiration. Spring is coming and soon the first question will be answered, which might just provide the momentum to begin to tackle the rest. Have we found the lost Intruder?

Sixteen

...QUELL URGENT DESIRES TO DIE

Winter 2015

Having served for ten years on active duty—including flying in combat—in the United States Navy did not make me eligible for a pension or stipend of any sort. It did not allow for tax-free shopping privileges at the on-base Commissary for groceries or the Navy Exchange for sundry goods. It did not provide for medical or dental benefits. It did not even permit access to the Naval Air Station to show my son the few buildings still standing where I used to work. All that remained to show for a decade of service to my country were boxed up medals, plaques, awards, my memories, and—the only perk that honestly lasts forever—friendships tempered by the steely combat of a long-misplaced youth.

On January 15, 2015, I had a once in a lifetime opportunity to turn back the clock a quarter century, if only temporarily; if only in fantasy. I had a chance to prowl decks and levels, over knee-knockers, through passageways, and up and down ladders; to enter ship's spaces not seen in more than twenty years. It was in these rooms that the seeds of serious first thoughts on my own mortality were planted. Today, I would step aboard U.S.S. *Ranger* a final time before she left Puget Sound Navy Shipyard without flags or fanfare, absent the traditional white uniformed sailors manning her rails. With boilers silent and cold under the perpetual darkness of a forgotten warrior, *Ranger* was making ready to be towed away to be rendered for scrap. For the morning, though, she was ours to share with former comrades in arms, to temporarily join physical reality with memory.

One day and two dozen years earlier, I had launched from *Ranger's* deck in the predawn Persian Gulf blackness into the unknowns of first combat. In certain ways, it marked the launch of the rest of my life—I was not the same person after that morning. But that was the past, worthy of reflection only when surrounded by old squadron-mates, and not as the topic of a moribund séance in a solitary mind. To have this opportunity presented at such a time in my life bared its teeth at coincidence, challenging life with the stubborn insistence of a young man's, and an old salt of a warship's, denial of fate. The invitation to the *Ranger* for a final visit arrived with the same puzzle-perfect fit as so many other meaningful coincidences experienced over the past year of searching for the lost Intruder, a forgotten jet that had flown from *Ranger's* moth-balled deck by a man now reliant on battery powered wires in his brain. It was another in a string of what I had come to consider omens of success. The alternate explanation was the mere illusion of prophecy, perhaps inspired by the twin influences of an active imagination and a bit of stray voltage. I smiled.

There was Jared's visit to the orthodontist, where fate brought Richard, the crew chief of the search and rescue helicopter that plucked Denby Starling and Chris Eagle from the frigid strait. And there was the surprise newspaper short take following the final article in the *Whidbey News-Times* concerning the lost Intruder in 1990, inviting library patrons to an *Andrea Doria* china display courtesy of Peter Hunt. Was the involvement of diver Paul Hangartner, the once-upon-a-time A-6 airframes mechanic, just another odd coincidence, adding a second former Attack Squadron 145 member as a critical member of the lost Intruder team? Even the happenstance way the project had come together raised an eyebrow: Ben and the MDS divers volunteering to help, the use of Craig Thorngren's ROV. Maybe I was cherry-picking situations, assigning meaning where the coincidence was not even especially noteworthy. I didn't care. Each event felt special. Each made me believe that we were destined to find 510, despite not having a clue as to why that should—never mind could—be possible.

Ross Wilhelm, a friend and former Attack Squadron 145 bombardier/navigator, had called me on the Monday before the tour with the *Ranger* invitation. I thought about my Navy flight log book and the three entries listing the aircraft flown from the *Ranger's* deck as A-6 bureau number 159572. *Ranger* was decommissioned in 1993. She had fought off the blow torches and wrecking yard cranes for 22 years. Why scrap her now? The more I thought about it, though, the more appropriate it became. Everything goes away, everybody dies—that's just the way things are. Maybe it was better to grapple with this fact in the close combat of reality than to push it off until it could no longer be ignored; before life's events intercepted a tired mind's fantasies in an ambush by the truth.

In a strange way, the discovery of the lost Intruder would bring meaning to the loss of *Ranger*. No one was going to raise 510, there would be no one tearing the old bird apart or scattering her ashes in a fragmentation of memory. Beneath Rosario Strait's protection of ripping current and darkly frigid waters, with giant ships steaming 240 feet overhead, the lost Intruder would rest in perpetual homage to *Ranger* and all who had left the journey early. The contact had to be the missing A-6. Otherwise, what did it all mean? What would be the point of anything?

In a clarity of intuition, I reached back to life before brain surgery to regain a sense of perspective. Logically, it made no difference. The contact could be the lost Intruder, indeed, but ultimately what did it matter whether it was the old A-6? It would make no substantive difference to any aspect of my—or anybody's—life. In many ways, in most ways, the lost Intruder had already served its purpose and now reclaimed its rightful place in the fading permanence of memory. It was what the lost Intruder represented that carried meaning, the purpose of the fight, the trophy of the game, but only if seen in tandem with the realization that winning or losing really didn't matter. There was no longer any inherent vitality to the lost Intruder as a physical entity. That satisfied my philosophical and spiritual curiosity, but not my human yearning to know. It didn't matter either way. August's hard fought and sought-after contact was the missing A-6; it had to be.

Tugg Thompson was one of those old friends traveling to Bremerton to say goodbye, a friend who fortunately had retired from the Navy and still had an identification card and access to Navy facilities. We had both been pilots when on active duty, but I would submit today by sitting in the right seat of his silver Accord, allowing him to do the driving onto the Navy base. Jared sat in the back, skipping school for a lesson in history from has-been shipmates, given on a ship that hadn't sailed in decades. As we drove onto the Keystone Ferry to travel from Whidbey Island to Port Townsend, Tugg's descriptions of past victories and foibles were unrelenting in their energy. He was a tenacious energizer bunny with a heart as big as his enthusiasm for flight, with every word threatening to spin out of control with a fiery clap of his hands. It would be a lesson for Jared not available in a dozen weeks or years of school.

The ferry rolled in the unseen swell of a wintry, still-dark, pre-dawn. Once in Port Townsend, it would be a one-hour drive to the Bremerton Naval Shipyard, where Ross Wilhelm would meet us for the tour. We would need to be off *Ranger* ninety minutes later, as she was slated to leave Bremerton in a month for the long haul around the Cape of Magellan, and then north to Brownsville, Texas. For a ship to die on dry ground took a great deal of preparation indeed.

A half dozen other squadron mates had opted to regret the invite, stating honestly, if imperfectly, that the event would be too sad. This, I could not fully understand. I had a vague sense of how a soft melancholy might threaten, but to have tales of the past intrude on the present, to manifest themselves through real emotion? No. To me, it was a celebration of a long-closed chapter, one so distant through the ravages of time, so alien to today's reality, that it was difficult to quite believe that the memories were real. Would it have been better to have *Ranger* slowly rust away pier-side without urgency or reason? Would it have been happier to know that *Ranger*, bereft of visitors or mission, would slowly flake into obscurity? Wasn't the anonymity of the scrap pile just the sort of tidy closure that so many seemed to be searching for in life?

My mood was far from sad when we got to *Ranger*. Walking her passageways was energizing as I eagerly peered into each darkened space for a glimpse of the familiar. The view was not disappointing: it was as if movers had come for her furniture and wall hangings, but left all else unmolested.

Walking the corridors, which had always been bare, it looked the same as entering the darkened ship after a night of liberty: mostly quiet, but with the jarring yells of revelers always threatening. Or, was it more akin to the walk from midnight rations, "mid-rats," to the ready room during the rolling night, standing the 15-minute alert, heavy flight gear hanging loosely on an uncaring young frame, eager for the urgency of a surprise launch; something critically important; a mission. It was both, and it was neither. It was real.

I left *Ranger* seeing and sharing with Jared far more than expected, feeling pretty damn good, without a hint of sadness. I asked Tugg on the return drive how he felt and he agreed. It was only once we were in the car on our way back to the ferry to Whidbey Island that Tugg started to run out of stories. But he had a rather hefty surprise still in store.

"Pete, I really think you should send off another FOIA request for the second JAG Investigation." Tugg dropped the bomb he must have been taking all day to arm.

I turned to face him as he deadpanned, looking straight ahead with hands on the wheel. "What the hell are you talking about?" I asked, slightly irritated. Was he mocking my preoccupation with the lost Intruder?

"Well, you asked me to check my logbook to see if I had flown 510 in any other squadrons. I flew it a bunch in the early 1980s in Japan." Tugg enjoyed a good story. "Good" meant long.

"Yes, I did." Patience, I thought, and asked, again, "What the hell are you talking about?"

Tugg paused before answering. "Sometime between 1983 and 1985, Attack Squadron 115 was on a training detachment to Cubi Point." Cubi Point was the name of the U.S. Navy airfield located on the Subic Bay Naval Base. Subic Bay was on the west coast of Luzon Island in the Philippines.

"There was a scheduled multi-plane night training flight to Tabones Rock, a practice bombing range. One of the crews was in their first run of the evening, a high-speed, low-level straight path attack." Tugg hesitated as a car slowed in front of us. A straight path attack was a systems delivery, a bombing run that relied entirely on aircraft systems—the inertial guidance, radar, laser, and computer—instead of visual cues to find and prosecute the target.

"Something must have distracted the crew because they picked up a slight descent but didn't notice. I have no idea why they didn't respond to the radar altimeter's warning, but the jet kept descending until it literally bounced off the water, like skipping a stone."

My jaw dropped. This was almost unbelievable. To skip an A-6 off the water without producing a tumbling fireball would require an exceedingly low rate of descent and a perfectly wings level attitude. Still, I suppose it was possible.

Tugg continued, "The crew claimed that the fuel in one of their drop tanks exploded midair, but the evidence pointed to the jet bouncing off the water. I don't know this for sure, but I'd bet that it was the same bird that Denby and Chris ejected from in 1989. I'd bet it was 510."

I considered the timing. My research already indicated that 510 had indeed been assigned to Attack Squadron 115 and the USS *Midway* during that time frame. There were a dozen A-6s in a squadron, making it at least a one in twelve chance that the jet that skipped off the water was 510. My thoughts raced.

"The crew knew that whatever caused the thump, it was really bad. They went back to Cubi Point to land, and then during the post flight inspection discovered severe damage to the airframe. Both engines had been pushed off their structural mounts. The only things holding the engines on to the fuselage were the fiberglass engine cowlings." Tugg paused before turning right on the road leading to the Port Townsend ferry.

"The intermediate maintenance facility in Japan worked on the jet for a long time. They had a reputation for outstanding quality repairs, really meticulous workmanship, but who knows what kind of structural damage had been done." Tugg let me digest the information.

It made sense. It was almost too neat, close to sounding contrived. Five-ten had been a maintenance nightmare during its tenure in Attack Squadron 145. Some of the squadron aviators, notably Denby Starling, had even started referring to the jet as "Christine" after the demonically possessed killer car from the Stephen King novel. The Tabones Rock incident might just explain why.

If the airframe had been deformed in some unusual way, maybe the airflow and drag of flight put stress on certain individual components unforeseen in the aircraft's design. Perhaps the problem could not be reproduced on the ground. If the airframe started to bend slightly in flight, possibly twisting just far enough out of normal alignment to put pressure on a hydraulic line, could this be the source of 510's problems? It made sense that a small deflection that ran the length or width of the fuselage might disproportionately impact fittings torqued to specification. Could the pressure induced by the lift and drag of flight be just enough to cause a hydraulic fitting to leak? Could it twist the fuselage sufficiently to stop the firing mechanism on a compressed nitrogen landing gear blow down cylinder from engaging? Was it reasonable to believe that the anomaly would disappear once the aircraft was slowed after landing when all would test normal?

It sounded perfectly logical to me, but would the theory hold muster to an aeronautical engineer or an airframes mechanic? Just about every A-6 that had been flown any length of time was bent to some degree, requiring rudder or flaperon trim to fly straight. The nature of 510's damage might be unique. The chances of an A-6 skipping off the water and returning to service seemed more far-fetched than the malfunctions caused by undetected or unrepaired damage.

I trusted Tugg completely, but still, I needed to hear from others to verify the details of his memory. It had been over thirty years, after all. Over the next several weeks his story was loosely corroborated by several others who had been stationed on *Midway* at the time of the incident, including our Commanding Officer at the time of the ejection, Commander Russ Palsgrove. The problem was that no one could remember the names of the aircrew involved in the incident. Tugg thought that it might have been a crew he could only identify by call-sign, their squadron nickname. I submitted a FOIA request for the JAG Manual Investigation of the incident, hoping for an easy answer. My letter cast a wide net to account for errors of memory as to the time frame. The reply came a month later: there was no such JAG Investigation on file.

That got my attention. Three people onboard the USS *Midway* at the time of the incident remembered the mishap. I had to conclude that the event had occurred. The missing report might have been lost, but could something mildly sinister have caused the report to disappear?

Suddenly, it struck me that the Navy might not have wanted the missing A-6 to be found. What if somebody did not want the connection made between the lost Intruder and the Tabones Rock incident? Ridiculous, I thought, uneasily. I pushed conspiracy theories from my mind and considered how else I might be able to determine the bureau number of the A-6 involved in the water-skipping incident.

Five-ten's maintenance logbooks would resolve the issue, but an online search indicated that these records were disposed of four years after a squadron decommissioned. Every A-6 squadron had been decommissioned for at least seventeen years. The JAG Investigation maintenance records for 510 only went back to 1985. I didn't have high confidence in anyone's memory of a specific six-digit bureau number after thirty-plus years. Still, some sort of written documentation of the event must exist, even if it were only in an aviator's logbook.

I asked Tugg to continue canvassing his old squadron mates for information. I also contacted several Navy friends not seen in decades and conducted long distance email and phone interviews. One friend, a non-flying maintenance officer, seemed to think that 510 was the same jet that had been involved in a refueling fire in the Philippines. He seemed to remember that it might also have run off an unidentified runway somewhere in Asia before being repaired and returned to service. Such memories are by no means inconsistent with the Tabones Rock incident. These later mishaps could have been caused by damage incurred by 510 skipping off the water.

Did someone in the chain of command have something to hide? Was it even possible for one man to sabotage a U.S. Navy search and salvage mission? It sounded unbelievable.

Looking for the answer became as frustrating for me as dropping a quarter in the checkout line at the grocery store, knowing that it will highlight my struggle with Parkinson's. Enough questions have been raised about 510 that I feel compelled to try to find answers. The answers lie in the past like the quarter lies on the floor at the market: I will try to pick it up, but I know all too well that my lack of finger dexterity might make it a futile lesson in patience. At some point either another customer will pick it up for me, or I will tire of holding up the line and give up, opening myself to the awkward question of why I'm leaving something of value behind. If a stranger does not volunteer the answer, then 510's mystery will likely remain unsolved.

By the end of early spring, favorable tides finally allowed for a dive attempt to identify the contact. I arranged with the Vessel Traffic Service to conduct diving operations in the shipping lanes on April 28, 2015. With the forecast weather at the margins, I made the call to go. Dan Warter, Rob Wilson, and Paul Hangartner met me at the marina soon after first light. We had barely finished packing *Sea Hunt's* aft deck with heavy Megalodon rebreathers, emergency stage bottles, and stacks of high-tech gear when the wind started blowing. We left the marina and almost immediately encountered seas too rough to dive. *Sea Hunt* turned into Rosario Beach to try to wait out the storm. Two hours later, with our slack current window of opportunity gone, we canceled.

It was still a productive day, one that provided the opportunity to get to know the MDS divers better. Our conversation also added substantially to the anecdotal evidence regarding the Tabones Rock incident. I asked Paul if he remembered anything about 510 from his time in Attack Squadron 145 as an airframes mechanic. To my surprise, he confidently answered yes. He was quite sure that he had helped conduct an acceptance inspection of 510 during his final days in the Navy, around March of 1988. The reason he was certain was that not only had the jet come from Japan, which didn't happen often, but this special inspection included the unusual requirement of a salt water intrusion assessment, a procedure unfamiliar to Paul. The situation was also memorable because the transferring A-6 had been dissembled and shipped from Japan in giant crates, a process that took weeks. Ordinarily, a transferring A-6 would refuel from an Air Force tanker during the long overseas leg, enabling the reassignment of the aircraft in a matter of days.

During the acceptance inspection, the armored steel plating bolted below the engines to protect the crew from enemy ground fire was removed. Much to Paul's surprise, only the four corner bolts were installed correctly—the remaining bolts were not screwed in at all but were held in their holes by some sort of putty-like compound. Something had caused significant, perhaps unique, damage to the belly area of 510 that could not be adequately repaired with the parts or capabilities available in Japan. Coupled with the need for a salt water intrusion inspection, it was compelling evidence that 510 might have been the same A-6 that skipped off the water in the Tabones Rock incident. The MDS team left the marina for home after vowing to try again.

I knew that weather cancellations were to be expected, and that didn't bother me unduly. But not all was well on a personal front. I was floundering, wallowing in the early stages of an increasingly apparent deep depression. For the first few months after the operations, I had ascribed the feeling to post-surgery fatigue, but it did not abate as the incisions healed into scars. Instead, it got much worse. By May, I found myself in a desperately apathetic struggle to do anything. I spent hours pacing, too tired and distracted to even pick up a book and read. It was a constant low-key battle that should have been easier to overcome than the pre-surgery challenges. But it wasn't.

I started drinking heavily, not every night, but damn close to it. I put on weight until reaching the heaviest I have ever been; then I added five more pounds. It was the opposite of my dichotomy in appearance and mood of six months earlier when I had been gauntly thin and looked like shit, but was happy. Now I looked distressingly normal, overweight and all, but I was profoundly sad. Plagued by persistent exhaustion and listlessness, I was being drawn into a downward spiral. I pushed away friends and acquaintances in what must have appeared to be a self-induced sabotage of isolation. Once disengaged, it seemed nearly impossible to return to where I had been socially only a month before. I began experiencing huge mood swings before reaching the bottom, marked by that most insidious of dangers—self-pity.

Daily survival depended on my ability to control mind and body. With great effort, I could fake an approximation of normal for hours, but would then inevitably collapse in shrunken despair. In a glimpse of understanding, I realized that my extant challenge was every bit as severe as last year's, maybe more so, only now, it was not visible to others. Where it only felt as if I was completely alone a year ago, now I really was isolated.

I saw depression as a sign of weakness, a reason to be ashamed. There were rational grounds for the depression. My body still could not produce sufficient Dopamine. Intense daytime tiredness is all too real for those with Parkinson's. When a sleep attack hits, I can't even talk to people, can't bring myself to form words, the utterly bone-tired sensation is so severe. Nothing came easy. Just rising from a chair to complete the simplest of chores could be a monumental task, but nobody could understand why. I looked normal, but I wasn't. Every single day, confronted with the same challenges as if they were new, a nightmarishly repeating Ground Hog Day of panicked loneliness.

I had forgotten how liberating it was to exhibit overt symptoms, to not feel a childish compulsion to attempt to hide my real self. Every movement was a chore—getting dressed, retrieving money or credit cards from a wallet, sitting straight with head unbowed while driving—it was the same nagging plague of nuisance faced five years before DBS surgery. It was double jeopardy, and even though I told no one, it was intensely shameful to know that I had uttered those self-defeating words of failure in the solitude of my mind: "It isn't fair."

I had created in my mind a fantasy vision of how life had been five years ago, forgetting the liberation of my soul when at last my Parkinson's symptoms were too obvious to ignore. When one looks healthy, but the slightest movement is painful and takes tremendous effort, it's a never-ending test to keep active. But when it is apparent to observers that something is desperately wrong with one's body, when it is all hanging out for the world to see, then the solitary struggle is just as often applauded as misunderstood. And everyone stands straighter to applause. Without obvious physical symptoms, the lapses indicating abnormality—keeping the line waiting at the checkout counter, not engaging in conversation, the blank look on my face, the inability to legibly write even my own name—are viewed as rudeness or laziness.

The Lost Intruder

And now, knowing that I was essentially light years ahead physically from where I had been six months earlier, the sleep attacks were accompanied by a panicked anxiety that I had much to do, but couldn't get motivated to complete the simplest chore. I was trapped in a horrible combination of frozen will and body.

Setting me back five years would replicate the hardships as well as the positives, but this time, none of it would be new. What worked last time to get me through might not serve me well the second time around. In some respects, it was the cruel surprises, the disease's unknowns, that had kept me active before DBS surgery.

I tried to reassure myself that nothing stays the same; that the progression of the disease, this time, would not be a carbon copy repeat, but would surely hold its own unique tests. I would be older too, with both kids out of the house, which could be good or bad. In a few short months, my mood had soured, and my sense of inner strength had evaporated. The seduction of easy, lazy living was tempting me and winning. The disease was winning.

At the depths of emotional despair, two things oriented my path forward, forcing me to move. I was fueled by a powerful curiosity, by a physical need to identify last August's sonar contact. This needed to be done before being able to fully consider what came next. It was also comforting to know that yesterday really didn't matter, that change was mine to make when ready. I just needed to keep trying. But I didn't have much time. The downward vortex was tightening.

We needed eyes on the contact for positive identification. Be careful what you wish for; you might just get it.

Seventeen

The battle won with the knowledge...

Spring/Summer 2015

The May 12 slack tide was predicted to be barely long enough to find and identify the contact before the currents grew too swift for underwater work. We had less than ninety minutes to pinpoint the faint ripple on the DownVision, drop the weighted downline at a drift corrected fix, and only then have the divers gear up and descend. Working in the middle of the busy shipping lanes didn't make things any easier. We would need to warn off approaching vessels for the duration of the divers' twenty minutes on the bottom, as well as their required hour of decompression. The MDS team was well trained, experienced, and professional, but make no mistake—diving 248 feet deep in open water is extremely dangerous; there is precious little room for error. Not having put technical divers over *Sea Hunt's* side in almost a year, and never to this depth, I was unsure of my capabilities, particularly given my pre-surgery lapses in performance. Barely hidden doubts still crowded the shadows of my mind, subtle fears that I would somehow fuck it up.

The long winter months had tested my patience, inflating the contact's importance disproportionate to reality. Completing this final task, getting eyes on the promising sonar return, was critical to understanding the events of the past year, which was the key, or so I thought, to moving my life forward. My fascination with the contact verged on the iconic. Towfish side-scan pictures stood faithful watch over me during the cold, dark months, pinned like sentinels above my desk. Dragonfly DownVision images haunted my final moments of wakefulness before sleep. Simultaneously, pre-surgery memories of last year's search began to grow fuzzy, as if I was bearing witness to events in someone else's life.

Sleep didn't come easy the night before the dive. Finally accepting wakefulness at 4:00 am, I picked up my smartphone and went online to check the automated winds at the Smith Island reporting station. It was blowing eighteen knots out of the west, more than enough to build a ground swell and generate whitecaps. I tabbed to the marine weather forecast and relaxed a little, taking heart that the winds were supposed to ease by 9:30 am. I left the house at first light.

After clearing *Sea Hunt's* aft deck to make room for dive gear, I went into the cabin to prepare coffee and start the engines. Ten minutes later, I moved *Sea Hunt* to the dock at the base of the marina ramp where it would be a shorter distance for the on load of dive gear.

I looked up to the crunching of gravel under tires as Rob Wilson's extended cab pickup truck turned off the paved road and entered the empty parking lot. The four doors opened simultaneously, but at first only Paul Hangartner and Dan Warter got out the back. A few seconds later, Josh Smith and Rob Wilson stepped out of the front. Ben had been unable to go, and Josh had agreed to assist as the deck hand for the day. He had injured his knee the prior dive season, and while fully healed, he was too rusty to join the other divers. Once the dive gear was loaded aboard, *Sea Hunt* got underway.

As we motored out through familiar waters, vague memories of the previous summer threatened from the subconscious; it was not just the towfish now at risk. The lives of the three divers were my responsibility, three men I had come to consider friends. I thought back to relinquishing the helm to Ben; to dragging the towfish on the bottom; to my out-of-control drive to keep trying. There was no balance. It's different now, I tried to reassure myself, all the while asking, "But can I now handle the pressure if something goes wrong?"

Clearing Deception Pass, a powerful swell abruptly lifted *Sea Hunt's* bow before falling in a loud crash, slinging loose objects across the cabin floor. Rob knelt in the galley to gather up the mess. As we passed to the north of Lawson Reef, the wind and current worked in tandem, pushing the water up the shoal from the flat bottom 250 feet below. Just when I was about to verbally question the sanity of the dive attempt, the wind shifted unexpectedly and then began to let up in the final mile before reaching our destination. I glanced at my watch; there was still a full hour before slack tide.

I keyed the microphone on the VHF radio and advised Seattle Traffic that we were approaching the shipping lanes for our prearranged four-hour block. Seattle Traffic acknowledged our check-in call, instructed us to contact them once divers entered the water, and then again when they were safely back on deck. The bow swung lazily as I steered the final hundred yards to the contact. *Sea Hunt* turned uncomfortably broadside to the three-foot swells, then rolled sloppily over the top of each wave. Impatient for the familiar DownVision image to appear on the Dragonfly display, I worked the throttles in opposite directions to hasten our slow-motion turn.

The current, wind, and waves conspired to make an accurate drop of the weighted downline a challenge. On the first attempt, Paul released the anchor on my mark while still up current from the contact, but by the time the 25-pound Danforth hit bottom, we had drifted too far away. I maneuvered the boat to pick up the downline, Paul pulled in the buoy, and we took turns hauling in the heavy anchor. While I jockeyed *Sea Hunt* into position for another try, Josh shackled a 20-pound collapsible grapnel hook onto the anchor chain for extra weight.

The second drop fared better. After making several runs close aboard the cluster of two floats, it appeared that the contact was on a 010-degree heading from the stationary downline, less than fifty feet from where the anchor rested on the flat bottom. Fifty feet was within a reasonable search range for the divers. The MDS team started to gear up.

The waves had noticeably subsided by the time the divers were ready to don their rebreathers. I steered *Sea Hunt* for three passes up current of the floating buoy-balls, dropping a single diver off on each run to assemble with the rest of the team on the surface. The three men then disappeared below the waterline, leaving Josh and me to guard their position against passing boats. As the divers vanished, I looked at my watch—9:44. The elapsed dive time would be our only indication of success, failure, or even tragedy until the divers surfaced. If the group came up any earlier than eighty minutes, it was probably a bad sign. Coming up substantially later might mean something much worse.

Although the wind had eased, it still required near constant throttle corrections to keep *Sea Hunt* within fifty yards of the buoys. We were not alone in the area, and it was important to stay close, like a sheep dog herding its flock. A one-hundred-foot scientific research vessel worked a fishing net a half mile to the west, perhaps counting local fish populations. Getting caught in an active fishing net was an especially gruesome possibility for a diver, and when the research vessel unexpectedly veered toward us, we successfully warned him away with a radio call. But not everyone on the water was so responsive.

Soon afterward, a small fishing boat raced full throttle directly toward *Sea Hunt* and the buoys, paying no heed to multiple radio calls and horn blasts. The speeding boat ignored the two large dive flags displayed from high above on *Sea Hunt's* fly bridge. At the last moment, I positioned *Sea Hunt* between the buoys and the speed boat while pointing to the dive flags, gesturing for them to stand off. The boat, with only one man and one woman aboard, continued straight ahead, apparently oblivious to our presence. Inside of fifty yards with barely sufficient room left to turn, the man at the wheel swerved just enough to avoid us. Like so many recreational boaters, they were ignorant of the most basic rules of the road on the water. They disappeared to the east, still running flat-out.

Twenty-five minutes after the divers submerged the red decompression buoy started to drift, separating from the stationary white downline ball. The divers had begun their ascent, signaled by the release of the weighted decompression buoy from the main downline at 120-feet deep. With the red buoy now free to move with the current the divers could complete their decompression effortlessly. The distance between the red and white buoys increased, and I jockeyed *Sea Hunt* into a comfortable position close aboard the drifting red ball. The slack tide was definitely over. The stationary white buoy grew smaller in the distance as we drifted north.

For the next twenty minutes, the only other visible boats were far off on the horizon. Then the calm was abruptly broken by radio static. Seattle Traffic warned of an ocean-going tugboat heading our way, eight miles out. Once within sight, the approaching Captain pleasantly answered my first radio call and detoured away from *Sea Hunt's* position. Just as the tug passed a quarter mile to the west, the divers broached the surface next to the red buoy ninety minutes after submerging. They must have found something, I thought while turning the stern toward the three men. With the boat's engines now in neutral, Josh stepped out onto the swim platform to lift the divers' unused emergency stage bottles into the boat. The divers then took turns climbing *Sea Hunt's* ladder. I bit my tongue and waited. Finally, Dan Warter, the first to exit the water, was free of his rebreather.

"It's definitely something manmade down there," Dan shouted over the noise of the idling engines. "It's not like anything I've ever seen."

Josh helped Rob Wilson get out of his dive gear. Rob gave me a curious look and shrugged his shoulders. I grew cautiously optimistic.

Then Paul Hangartner spoke up.

"That's not an A-6," He said confidently.

Just that quickly, the nine-month-old question was answered. Shit.

Whatever the contact was, however, it was clearly man-made. The downline had led the divers to within twenty feet of the sonar return despite the difficult anchor drop. They found the contact immediately, allowing Dan fifteen minutes to video the mystery wreck. The lost Intruder project had to end eventually but at the price of failure? I looked silently to the divers for a clue as they methodically removed their gear. The hell with it, it was still my decision. It wasn't over, not yet.

It took a careful review of Dan's underwater video and online research before we discovered the true nature of the contact. It was a tank container, a cylinder that looked like a giant propane tank used to ship liquid or gaseous goods. Because the tank was rounded, an external steel frame was required to square off its features so that it could be stacked with like-sized containers on the deck of a ship or barge. A three-foot piece of the external frame had been bent upward: this was my A-6 "probe." The tank container had been underwater for at least a decade.

Deterioration of the external frame, coupled with the rounded, hard return of the tank in the center, provided a sonar picture that was easily manipulated by an imagination that saw what it wanted to see. The tank was intact, with only its thermal melt-away coating, similar to a thick paint, eroded away in small patches. The thermal coating was applied to buy time to extinguish a fire before the volatile compound inside either escaped or exploded. The nature of the hazardous contents was a mystery. Josh helped the divers out of their gear while I steered *Sea Hunt* to retrieve the downline.

We cruised back to the marina in silence and a depressing absence of activity that took me the better part of a week to shake. The lost Intruder had controlled me when I needed the steady hand of even purpose; it had provided the rationale for a winning strategy to combat a critical phase of my disease. But where did that leave me now?

Strangely, I still had confidence in finding the A-6. There remained several smaller targets from the previous year's search that we had not yet investigated; each might yield a clue to discovering 510's final resting spot. For the remainder of the Spring, Rob and Paul identified these contacts, all of which turned out to be nothing but rocks. Ben, over committed between work and unrelated searches on Lake Washington, had no time to devote to the project in 2015. Finding the lost Intruder as an ambitiously professional undertaking was stalled, perhaps dead.

I toyed with the idea of buying a commercial sonar or maybe just renting one, but the question always came back to, why? What had been urgent to me a year earlier held little attraction now. I refused to allow 510 to become an albatross around my neck. There are no quick, easy answers to anything. To look for one through the surrogate of the lost Intruder project devalued what the experience had come to mean to me.

Spring faded into summer, and I realized that it was time to either step up or off. To keep the project moving forward, I decided to continue the search with just the Dragonfly. It was a long-shot tool for the job, to put it mildly.

& August 2015 &

Sea Hunt works the small grid, exploring the gradually sloping bottom, smoothly uniform in its gentle rise from 220 to 200-feet deep. I look for any anomaly, a single bump that might provide a clue. Turning the helm 180 degrees to starboard, *Sea Hunt* runs parallel to the track line left on the Dragonfly display from the boat's previous circuit. Despite the 2-knot current, *Sea Hunt* stays close enough to my limit of 50 feet from the dashed track line. When the pattern exceeds 50 feet, becoming too wide, either due to the wind, current, or helmsmen error, I steer the next pass between the two most recent track lines left from previous runs on the chart plotter. By splitting down the middle what has broadened, to say, an 80-foot spacing, the remaining gap shrinks to only 40 feet. This should be a close enough interval for the DownVision to detect some sort of return if the sonar beam crosses an A-6 sized target. At least that's my hope.

My Parkinson's induced difficulties in maintaining control of the helm are but a hazy memory, one that grows increasingly dim with alarming speed. Keeping one eye on the Dragonfly display, I steer *Sea Hunt* the third time around the oblong pattern. I learn to never look away from the Dragonfly's screen for long—you never know what you might find.

Of all the time spent alone in the summer of 2015, and there is much, the days on the water exploring are the most fulfilling. They are meditative. It is also all that is left of the lost Intruder project, with *Sea Hunt's* deck now starkly empty. No winch leaking hydraulic fluid dominates the cockpit. No ROV tether cable is strewn about, leaving no place left to stand. The cabin seems barren, deserted. No computer monitors are cluttering the surfaces, or extension cords lining the deck. No conversations, no other people. It is only me, wondering and wandering as *Sea Hunt* combs the bottom. Just me, the chart plotter, and the DownVision. It's much as I first envisioned the search, an effort of solitude, probably a fruitless, perhaps even a pointless, endeavor. The project offered focus and distraction in 2014, but both were fleeting enticements. Both mean little with the temporary easing of the worst of Parkinson's physical symptoms.

Still, I continue to look. The project offered a continuity to a past life unencumbered by Parkinson's; it restored a purposeful vitality to my life.

The Lost Intruder

The active search resonates within me, bridging the gap between past and future. It strikes me that it is important to keep looking; it is the finding that has started to lose meaning. *Sea Hunt* and I spend one to two days a week of favorable currents and weather crisscrossing the area most likely to be 510's hiding spot. This is the sector consistent with what I have come to consider the "three hard pieces" of data. I've designated the overlapping sector derived from these hard pieces of data my new high probability search area. Searching this relatively small grid gives me a new feeling, one that I am not alone.

The first piece of "hard data" is 510's exact position during ejection. This is the most reliable evidence available. It is derived directly from the air traffic control radar logs and matches up perfectly with the time-stamped radio transcripts. I then expand 510's course after ejection into a 20-degree cone to the right to account for heading deviations to starboard after the crew ejected; Ray Roberts testified in his statement that 510 dipped its right wing down. There is still the question of just how far the A-6 flew before crashing, but the intersection of the other hard data points with 510's approximate course line yields a reasonably small patch of ocean, one that just might be searchable with only the Dragonfly.

The second set of hard data is based on the current, and to a lesser extent, the wind. Working backward with the tidal information from the day of the ejection, I reverse the effects of the currents and the wind over time. Backtracking the floating debris field discovered by the Coast Guard Cutter *Point Doran* should lead directly to the crash site, but only if I can replicate the effects of 80 minutes of current and wind correctly from one blustery afternoon 26 years earlier. After hundreds of hours of observing, I have become far more knowledgeable about the eccentricities of Rosario Strait's currents. The effects of the wind on a small profile object bobbing in the swell are not as pronounced as I once thought. Contrary to my first inclination, the tidal flow in the area is remarkably consistent. The exception is during periods of harsh weather when storms of extreme barometric pressure can cause the currents near the crash site to be unpredictable. This is unusual, however, and the day in question did not have an excessively unruly barometer.

I have come to consider the third category of hard evidence with a healthy degree of skepticism. It is here that I depart from my previous practice of treating more information as better. The key to finding 510, I have come to believe, is through limiting individual pieces of evidence. I use what information rings true, disregarding isolated data points that don't feel right as I attempt to intuitively reconcile the complexities of many variables. When one focuses on everything, the reality can be that the actual focus is on nothing at all.

Eyewitness azimuth reports are reasonably dependable, relying as they do on the alignment of landmarks on shore. Accordingly, bearing information from a variety of sources is included in my new modeling. Judging distance over water, however, is inherently problematic, and I discount the aircrew distance estimates entirely.

In retrospect, this was an obvious shortcoming of both of my earlier theories. Chris Eagle's ¼ to ½ nautical mile; Ray Roberts's ¾ to 1 mile; Denby Starling's several miles—they simply can't all be correct. Furthermore, none of the distant estimates are consistent with the far more reliable data of the location of the floating debris field. The more time spent on the water attempting to estimate distances, the more apparent it becomes how difficult it is to do so with any accuracy. Add in the stress of an ejection, and it is a fool's errand to put much stock in any of the distance estimates. As the *Salvor* did before me, I ignore them all. With all eyewitness distance estimates disregarded, every other observation magically fits a common narrative.

For example, the elapsed time from ejection until Chris Eagle saw the A-6 bubbling in the water now makes sense. This can't be more than twenty seconds given the ejection seat's timeline before parachute opening. Starting with the precise ejection point from the air traffic control radar logs, twenty seconds cannot be reconciled with Chris Eagle's or Ray Robert's distance estimates from shore. Of course, this only became apparent with the acquisition of the JAG Investigation. Trying to square the different distance estimates with all the other clues was one giant red herring.

The new high probability search area runs along the prevailing current line leading to the floating debris field. Working backward from the debris field is common sense, but the trick is correctly estimating the current's actual direction and magnitude, no small feat in the rapidly shifting flows of Rosario Strait. My new theory's water impact site is now slightly less than one mile from the ejection point, which is consistent with the flight simulator recreations of the accident in 1989. Chris Eagle's alignment of 510's bubbles with Lopez Island also now make sense, at least from the perspective of his sight picture while floating down in a parachute from 3,500 feet. By charting the centers of the wingman's orbiting radar skin-paints, the ejected aircrew's approximate positions can be determined as they descended to the water. These orbits are consistent with the prevailing wind at the time of ejection. They also align with the spot where Richard, the search and rescue crew chief, stated that the helicopter found Starling and Eagle.

Even a piece of previously ignored evidence now fits in with my new theory. A tower air traffic controller reported that 510 impacted the water to the southeast of Lopez Island. This eyewitness account of the ejection point, disregarded earlier because it conflicted with all three aircrew distance estimates, goes right down the middle of the new high probability zone. Except for the erroneous air traffic control interpretation of the radar logs—which has already been proven false—all the evidence is now consistent. But only after all distance estimates are ignored.

The new search sector is oblong with a total area of about ½ square nautical mile. This is a small enough area to explore using the DownVision sonar alone, but only because the sea floor is extremely flat. However, at about 210 feet, it is still deep. The winds on my chosen search days need to be dead calm, with nothing stronger than a modest current. It takes extraordinary patience and discipline to cover the entire sector during the few August days of perfect conditions. Judging by the appearance of the tank container's sonar return on the Dragonfly, I am confident that 510 will be revealed as a tiny, but distinct, sonar ripple. But it will only be seen by staring so hard at the DownVision as to practically not see the sonar contour; it almost needs to be felt. It is a meditation, with the dead-calm water a chakra, my centered source of applied energy pulsing below, of being "one" with the sonar. I know this sounds flighty-new-age and weird, but it is true. Maintaining this concentrated awareness for three to four hours at a time leaves me exhausted.

By mid-August, I have completed five search days in only the flattest of seas and mildest current. After twenty hours of staring at the DownVision, the entire high probability zone has been covered. Having to steer the boat precisely, with one eye always on the Dragonfly, while eating and drinking, even while urinating in a bottle, has me laughing softly at what I must look like. And at the apparent futility of the task. What utter madness, I think, of finding an A-6 sized contact with only the DownVision, a target that has been corroding in 210 feet of salt water for 26 years; a jet that the Navy had searched for with four ships, spending tens of thousands of dollars in the process. I know all too well that the Dragonfly is the wrong tool for the job. Still, my faith in finding 510 is inexplicably soothing. My final sprint to find the lost Intruder is intense and exciting, but it is also profoundly relaxing, bringing with it a newfound inner peace.

The days of work are not in vain. With the search of the new high probability area complete, I have found several clusters of minuscule contacts, all within a hundred feet of one another. The location of the sonar returns conforms to within .1 nautical miles of the center of my new model. What are the chances, I ask myself? Somehow, I know that they are excellent.

The most likely cluster of sonar returns appears to be 40 to 50-feet long, which corresponds to the rough dimensions of an A-6. It sits, at its highest, just 3 to 4 feet off the bottom, but it is tough to tell, as this precise a measurement blends with the DownVision's background clutter at the contact's depth. The DownVision's split screen option, offering a four-fold magnification of the bottom, is a great help in identifying the tiny black and white bulge, but a 4-foot anomaly is still almost impossible to see. Four feet is not an unrealistic height for an A-6 fuselage after two and a half decades of decay and collapse.

There is a second, smaller grouping of sonar returns that appear to be draped in a fishing net, making its size difficult to judge. An alternate explanation is that the DownVision could be depicting the sonar returns of schools of small fish, "bait balls," that hover at the top of an artificial reef. Both clusters lie in the middle of the twenty-foot rise in the bottom

The Lost Intruder

contour. If 510's wings broke off at water impact, is it conceivable that they both landed on the sea floor within a hundred feet of the fuselage? I can think of no reason why not.

It is a paradox that as less time is left to us on earth, most things become less urgent. I know that there will probably be only a handful of opportunities remaining to dive these new contacts in 2015. If the weather holds, we will know for certain if the contact is the lost Intruder, probably in a matter of weeks.

To gather physical evidence of the contact's composition, I decide to attempt to drag a grapnel anchor into the target. After rigging a steel grapnel hook with a three-foot length of heavy chain, and with the assistance of Jared and his friend Max O'Neil, we attempt to snag the hook into the contact.

Max is excited, and I overhear him say to Jared, "Wouldn't it be cool if we found the A-6 today!"

Jared, having already spent several days on *Sea Hunt* searching and finding nothing, tries to temper Max's enthusiasm. He answers realistically, "It's a real long shot. Really long."

Despite a reasonably mild tide, it is still damn hard to position *Sea Hunt* properly to align the grapnel with the tiny sonar return. The faint image seems to change location, not by much, maybe 20 to 30 feet, but just enough to be confusing. Time and again, I grow confident that the sonar return is narrow and about 40 feet long, only to have it suddenly vanish for several consecutive runs. Then it invariably reappears, clear as day, but only for a single scan. It is an anthill-sized target from where we gently roll on the still water 210 feet above.

On the fifth attempt of dropping, and then hauling back in, 250 feet of weighted line, the grapnel catches. It grabs hard enough that no combination of rapidly letting out line and pulling with *Sea Hunt* from the opposite direction will free the hook. I know from wreck diving experience in the distant past that this is an excellent sign. Usually, only man-made objects, like a shipwreck, refuse to relinquish a grapnel. I buoy the line, return to port, and email the MDS divers the positive news. Max is right; it would be cool.

We plan a tentative dive date of acceptable currents two weeks in the future. When the day finally comes, the weather is horrible, but the forecast indicates that perhaps it will improve. The wind and a heaving groundswell contribute to an inability to paint a consistent sonar return, or to successfully retrieve the grapnel left on the site, or to set a reliable downline. After three hours of trying, we lose the line running to the grapnel and cancel in frustration. I wonder if Rob and Paul will take another shot at diving the hard to pinpoint contact after spending so much time and effort. It would be understandable. The two men, having spent dozens of hours preparing and diving, have absolutely nothing concrete to show for their efforts.

Rob Wilson sees the setbacks as a sign, and makes the tongue in cheek comment, "The wreck is trying to warn us off— challenge accepted! Now we have to come back."

I chuckle at first, but on the motor back to the marina I remember 510's squadron nickname, "Christine." Does Christine want us gone? I try to laugh a bit louder, but I can't. I have that feeling again. We are not alone.

Dates of mild current and sufficient daylight are fast becoming rare as the season prepares to change. Rob and Paul agree on short notice to try again on Tuesday, September 22. The summer is winding down; the project is approaching the two-year mark. We badly need a win. And soon.

Eighteen

...THAT UNWINNABLE WAR LIES AHEAD

Autumn 2015

We leave the dock at 11:30 am with Rob Wilson's father, Jerry, and Chris Burgess, one of the friends who gave me the Dragonfly as a birthday gift eighteen months earlier, going along as deck hands. I search for emotional balance, an exceedingly scarce commodity of late. Post-surgery Parkinson's has not helped my spiritual equilibrium, but I'm finally coming to grips with my evolving situation.

There is evidence to suggest that impulse control issues can be caused by Parkinson's itself, or by either high doses of Levodopa or the Dopamine-mimicking drug Requip. Inquiring about impulse control issues had been a regular aspect of my neurological appointments pre-surgery, but I don't recall discussing the possible adverse effects of DBS itself on impulse control. DBS introduced three potentially life altering agents of change to my condition: the effects of the surgery itself, and acclimatization to both the hardware within my body and the constant electrical pulses being sent to my brain. Some of the after effects are fleeting, others take longer to dissipate, and possibly certain changes will become permanent. A year and a half after the procedure, I still occasionally experience an out of place feeling, of floundering, but it has gotten less severe over time.

There is also speculation that the DBS implanted electrodes might cause impulse control problems as well. Once discovered through an online search, the confused-hell of the past eight months suddenly makes sense. My excessive drinking, over-eating, and overdoing of most everything bad for me are suddenly explained. Depression—developed from a feeling of anomie, a free-floating listless isolation, of having the plane fly me—is the predictable result. I still have Parkinson's disease. It will continue to get worse. To function, I must still take high doses of Levodopa: 2,000 mg a day. I also tend to isolate myself, losing human connection, making me acutely susceptible to Parkinson's apathy and depression as well as addictive behavior.

Putting the guilt from my lapsing discipline in proper perspective helps me think logically about how to control these problems. I direct my tendency to obsess toward the missing A-6. I finally understand a critical aspect to my surviving 2014—the lost Intruder became a constructive outlet for my impulses. The project provided more than a morale boost; it also served a physiological purpose. Taking a lesson from pre-surgery, I take control and work to reduce my medications. I stop taking Requip and drop my daily dosage of Levodopa by 250 milligrams. I decrease the voltage on the pulse generators slightly. I sleep better, which makes the new drug regimen more effective.

It's a familiar pattern of interconnectedness between medication, exercise, sleep, and attitude. Dystonia is still my greatest physical threat and, fortunately, it does not return with the lower daily dose of Levodopa. The result is an ability to function on some afternoons without experiencing an apathetic exhaustion of futility. After eight months of uneasy confusion, I may have finally found my stride.

It's a beautiful September day on the water, and Rosario Strait unfolds before us with the shimmering beauty reserved for the last, failing days of summer. As usual, Rob and Paul are relaxed. They appear to have confidence in my ability. It is suddenly apparent how important this is to me, to be trusted by divers risking their lives for the love of exploration. I'm humbled and moved; we each have a critical role to play, and we act as a real team.

The Lost Intruder

During 2014, the lost Intruder project served as proof that my life was not just memories, that events twenty and thirty years earlier had indeed occurred, offering a sliver of stability. Ultimately, it proved my sanity, maybe even relevance, if only to me. To continue the search now with last year's hell-bent intensity would prove nothing. Conversely, to give up looking altogether would be to deny my initial curiosity. It would be admitting defeat. The month-long search with the Dragonfly feels the right balance.

The struggle to redefine my new, battery-powered life marches forward along with the evolving lost Intruder project. It is intuitively important to me to continue in some fashion with the project, but it is also equally imperative to move on to a new challenge. Otherwise, I run the risk of letting the plane fly me. It would be a disservice to the memory of the lost Intruder to allow her to manipulate me decades after I sat with anticipation at her controls. It was my hands and feet that commanded 510's stick and rudder to maneuver through the open skies back then. I must continue to direct my life to move in a similar manner now, with a clear direction and strategy.

The single constant of the last eighteen months lies 210 feet below where I sit at *Sea Hunt's* helm. *Sea Hunt* has reached the chart plotter's marked waypoints that define what I'm increasingly confident is 510. I know this, somehow, despite the odds against finding such a small contact with only research, the Dragonfly depth sounder, and persistence. Like running across an old friend not seen in decades, the settled sureness of this knowledge rests comfortably within me. Still, as with all aspects of life, there is a sliver of doubt.

Rob Wilson and Paul Hangartner get suited up. Once again, I'm awed by the complexity of their dive gear: heavy, compact rebreathers, two or even three stage bottles clipped to each diver's side. In one respect, it is absolute insanity. I smile inwardly, recognizing the common excuse of those not willing to risk it all for exploration's sake. I now consider myself within these general ranks, but I do remember what it was like to attack life as if there's no tomorrow, understanding that ultimately there really is none. The MDS divers inspire. The fact that they trust me with their surface support—me, who just a year earlier could not even speak coherently by afternoon—is overwhelming. Those rare people who can adapt and roll with the punches have always impressed me. I turn red, realizing that this also refers to me. The hell with it—I'm starting to be content again being me.

One at a time, Rob and Paul clip on the last of their equipment, the emergency stage bottles, and step off backward into the water. Each man swims slowly, with carefully metered effort, toward the downline buoy. They take a few minutes to compose themselves on the surface and then descend as a team. For the next 67 minutes, the only evidence of their existence will be the intermittent stream of tiny bubbles from a vented dry suit or cleared mask. I grow nervous. This is not worth anyone's life. In the same frame of thought comes the realization that, bullshit, it is indeed worth life: by forcing an examination of how and why one lives, it contributes to the possibility of a peaceful death when the time does come. I wait for the pair to surface, anticipating confirmation of what I already know. Still reeling from the hubris of the prior year, I refuse to say the words, but I am confident. We are not alone.

I've never dived with a rebreather, but it's not difficult to imagine the divers' underwater progress. I'm familiar with the theory and have watched the MDS divers work underwater in person and on video. Everything is done with painstaking slowness, nothing is rushed. The divers descend at a leisurely pace to the bottom, regularly pumping gas into their wings style buoyancy compensators to counter the pull of the crushing depth. Halting their plunge just feet above the mud, they are careful not to touch the 210-foot deep sea floor with an errant fin kick. Slowly, Rob snaps a thin nylon line from a guide reel to the downline chain, all the while maintaining perfect buoyancy. Paul attaches a flashing strobe light to the downline to use as a visual guide. I know that if it were me in their position, I would take a moment to touch the regulator attached to the stage bottle filled with bottom gas mix to ensure that it can be reached in case of an emergency. Then the divers set out.

Rob references the compass strapped to the back of his hand as he follows the estimated course from the anchor to the contact. The wreckage is too small, too dispersed to be sure of dropping directly onto the main debris field. The divers need an initial direction to swim to assist them in their time constrained search. They start to look, hovering above the sea floor, moving forward steadily with knees bent and fins high off the muddy bottom making gentle frog kicks.

Then my imagination goes blank. I have no idea what the divers will see before they reverse course on the nylon line to return to their starting point. When the divers do ascend, they will rise at no more than a foot every two seconds to their first decompression stop, probably at 120 feet. Many more decompression stops will follow.

I sit at the helm and scan the horizon. Once the downline buoys separate, the red one drifts with the current, marking the location of the decompressing divers below. I keep *Sea Hunt* close. I can't be of any help to the divers while they are underwater, but I can make damn sure that the boat is in position if something goes wrong. Providing surface support is both exciting and nerve-wracking. Rob witnessed my gradual deterioration in 2014, and Paul has seen me at the absolute worst, my lowest point when I surrendered the helm to Ben off Port Townsend. Once again, I'm awed by the pair's trust in me. Rob Wilson, Paul Hangartner, and Dan Warter have been the project divers. They are the real explorers. They take all the risk. I am pleasantly jealous, but also inspired and humbled that they are a part of the team.

In the days of flying the A-6, Intruder outfits had a uniquely important role on the aircraft carrier—the attack squadrons were the pointy end of America's spear. Everything else the carrier did was designed to support the Attack mission. I miss the days of being at the pointy end, but I now also take modest pride in providing support to the divers. Where "support" was once a position to be eschewed, now it feels right. It fits.

Finally, Rob and Paul surface. They both give me the "okay" sign overhead, and I visibly relax. Because of a minor equipment problem, the two divers did not make it to the primary target. But they didn't come back empty-handed. Somehow, Paul has managed to retrieve something manmade. Rob describes a thick silt bottom with awful visibility strewn with thin wires, the kind found under an aircraft's skin. My pulse races.

Paul gives his opinion. The fiberglass part he has recovered is a piece of air ducting of some sort. There are two small placards riveted to the encrusted object, and I break out an old dive knife to use in place of a dainty archeological tool to clean the growth from the writing. One of the plastic placards displays a part number that starts with "128"; the other shows a 1983 date of manufacture. The design appears to be aviation oriented, Rob says expertly, seamlessly reverting to his role as a Boeing aircraft maintenance technician. The unpainted fiberglass is in good shape except for the near uniform covering of marine growth. Unpainted means it came from the interior of something, protected from the corrosive salt air, but what? We are cautiously optimistic. After unloading the boat, I go home to research the part.

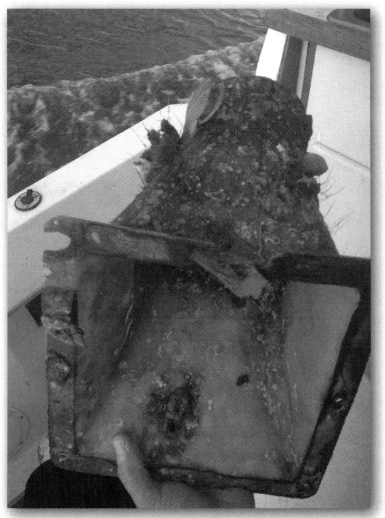

Fiberglass air ducting found by Paul Hangartner (photo courtesy of Rob Wilson).

The next morning the evidence starts to trickle in. Old shipmates have pulled together without even realizing the coordination of their actions. The squadron's former Maintenance Control Officer, Al Gonzales, confirms that the prefix of "128" indeed denotes a Grumman part number. Grumman made everything from jets to buses to mail trucks in the 1980s, however, so while this alone does not cinch things, it is an internal Grumman-made component. It did not randomly fall out of the interior of a passing jet, and it certainly did not fall off a mail truck.

Grumman placard on air duct (from the author's collection).

Rob finds a cutaway sketch of an A-6 online and identifies a component within the tail section that looks remarkably like the recovered ducting. I agree with his assessment. The artifact appears to be a piece of the aft equipment bay cooling ducting. It was installed in the empennage. If the Intruder's tail indeed broke off, as seemed to be indicated by the composition of floating debris picked up by the Cutter *Point Doran*, then the parts that sank should be close to the fuselage.

I send several photos and a brief explanation via Facebook message to one of the 1989 Squadron work center supervisors. John Edgren now lives in the Midwest. At the time of the accident, he was the supervisor of the

Part number and 1983 date of manufacture placard
on air duct (from the author's collection).

work center that repaired the A-6 cooling systems. Facebook has allowed me to instantly ask questions of any number of experts who had performed maintenance on 510. John replies in a matter of minutes, he is unequivocal—yes, the photos are of a piece of ducting from inside the tail of an A-6. It is the same part portrayed in the schematic. It was used to port ram air from outside the jet for electronics cooling in the aft equipment bay.

Ironically, John Edgren's work center also repaired and maintained the Intruder's ejection seats. I vaguely remember Denby Starling and Chris Eagle buying John a bottle of expensive liquor after surviving the ejection as a "we would have died if not for your work center" traditional gesture of thanks. The scores of moving parts in the two ejection seats had all worked as advertised. It is fitting that John is the one to corroborate the duct's origin, thereby indirectly fixing the position of where 510 hit the water and sank. It is surreal that we are approaching the end of our journey. It's difficult to keep up with answers that now come so sure and quick. It is hard to fathom that I had flown the A-6 Intruder that housed the duct. From an aircraft carrier. At night.

Still, it is circumstantial evidence; no one has yet laid eyes on the missing A-6 itself. Or what remains of the missing jet. The wires Rob found strewn along the bottom point to the site being near the center of the debris field. The distinct hard contact in the soft, mud bottom depicted on the Dragonfly is long enough to be the fuselage. It is only fifty feet away from where the duct was found. The manufacturing placard stamped with a 1983 production date places it in the correct timeframe. There has only been one Grumman-built aircraft that has crashed in the waters off N.A.S. Whidbey since 1983. No other military aircraft have been lost in this vicinity since 1983. I'm convinced. It's just a matter of time now before we gather irrefutable proof. I email the team a simple message:

"We got it."

The replies come quickly, first from Rob Wilson:

"*OUT-FUCKING-STANDING.*"

I grin at the disregard for niceties that is the hallmark of the serious wreck diver. Rob's email continues: "Going to be hard to get the smile off my face today. Congrats, Pete you did it. "

I'm grinning ear-to-ear now, not with last year's crooked attempt at a smile, but with a genuine look of pleasure. I reply:

"Correction—we did it!"

It isn't until Friday, October 16th that the currents allow for further exploration. Rob and Paul jump at the opportunity to try the site again. We also plan for a dive on Saturday, October 17th as a weather backup or for more mapping of the bottom. Dan Warter will join the other MDS divers on Saturday and plans to document the dive on video. Suddenly, there is no shortage of volunteer deckhands. A hunk of growth-encrusted fiberglass instills instant credibility in what was becoming a white-whale waste of time and effort. Perception jolts into reality, perhaps temporarily, more likely to stay.

The weather holds in a long Indian summer as the dive dates approach, easing what has been a growing fear—being forced to wait until spring for positive confirmation. As the one-year anniversary of DBS surgery approaches, I am finally flying the airplane. Energy starts to return to my body. I gradually begin to lose weight. I strive to work out every day, and for the most part, I do. Sleep still comes with difficulty, but I find that at least with the new regimen it is much-improved REM sleep. There are fewer sleep attacks and sudden bouts of anxiety. It all comes together in a rush. I sense happiness, not quite present yet, but lurking pleasantly somewhere around the corner. "Christine" remains fickle, however, and the lost Intruder gives up her secrets only after investing tremendous effort.

Rob and Paul have fabricated a concrete anchor by using a traffic cone as a form. The plan is to drop the sixty-pound cement weight fashioned to a semi-permanent buoy to use as the starting point for an archeological-type inspection grid. Rob and Paul have made two deep dives from shore at Mukilteo in the past week to practice deploying the numbered lines along cardinal directions. They will then be able to identify and mark as many items in the debris field as possible. Subsequent divers will be able to reference their location in the debris field by the closest color coded, numbered line.

We find the elusive sonar returns on the first run, drop the cement cone, and let out the downline. Rob and Paul prepare to dive. The divers spend 30 minutes on the bottom, for a total dive time of 110 minutes including decompression. Josh Smith, Chris Burgess, and I watch the drifting red buoy that marks their decompression line as the minutes slowly tick by. Finally, the two surface. I wait patiently as Paul hands up his stage bottles to Josh standing on the dive platform. I catch Rob's eye behind his dive mask as he comfortably floats behind Paul. Without letting his rebreather mouthpiece slip from between his teeth, Rob calmly gives me a thumbs up. I can see from the sparkle in his eyes that he is smiling.

I clap my hands a single time in excitement, and then abruptly turn and go into *Sea Hunt's* cabin. My face relaxes for a moment as I stare at the helm. I can't believe any of this. My shoulders slump forward with a deep exhale, and the strength leaves my legs for a moment. I bunch up my shirt sleeve and dab at my eyes. I take a deep breath.

Holy shit, I think, eyes wide; we did it.

We really did it.

I raise my eyes up to the cabin ceiling, easily beating Parkinson's pressure to bow my head. I won the battle, motherfucker. You lose this round. I tense up and give the emptiness beyond the ceiling a firm middle finger.

You lose.

The Lost Intruder

I return on deck thirty seconds later with a bottle of Champagne. Rob and Paul join us in a rough, wordless toast straight from the bottle before they have even climbed the dive ladder out of the water. The bottle is empty well before either man is out of his dive gear.

The divers have found a dense debris field. Some parts are identifiable by description as clearly from an aircraft, and almost certainly an A-6. There is no question in my mind that we have found the lost Intruder, but we still do not know if there is an area with more fully intact pieces of wreckage. We can't tell for sure if we have found the main part of the jet, or even if such a thing exists. Excited and happy nonetheless, we cruise back to port as the sun sets, ready to give it another shot the next day. We leave the buoy to mark the missing jet's position so that Rob and Paul can continue working their grid line on Saturday while Dan Warter films the evolution.

October 17, 2015: Preparing to depart the marina to video 510's wreckage. On the dock: Rob Wilson and Peter Hunt. On *Sea Hunt*: Paul Hangartner and Dan Warter (photo courtesy of Josh Smith).

The weather holds beautifully, but the dive will be running right up against sunset. This could be our final opportunity to locate the main wreckage for the year. Any nervousness felt by the crew is masked by excitement. We leave the marina at 2:30 pm. Sunset is 6:25. It will be awfully close, but things should go faster now that we have a buoy already marking the site.

There's just one problem. We arrive at the proper latitude and longitude coordinates and look for the commercial, white buoy placed at the site just 24-hours earlier. It's not there. There are almost no other boats on the water this late in the year, and we are a half a mile from the nearest shipping lane. What the hell happened to the buoy? Is Christine toying with us? The deckhands—Josh Smith and Dan Crookes—hurriedly fashion another downline with an extra Danforth anchor and chain I brought along for just such a contingency. The only rope we have is a 300-foot spool of floating polypropylene line, which will not make anything easier. I will need to be extra careful so as not to tangle the floating line in the boat's propellers while dropping the anchor and rigging the buoy, as well as when we retrieve the setup and haul it back into the boat. The polypropylene line causes the detachable decompression line buoy to float differently, counterintuitively, relative to the downline buoy. It gives no indication of when the current begins to calm. I tell the divers to suit up while we try to sort out the mess. Christine is here, I can sense her presence. We are closing in.

The current slows and the divers descend out of sight along the downline. Josh, Dan Crookes, and I wait, eyes scanning the water's surface as the sun lowers on the horizon. If all goes well underwater, sunset will not matter. *Sea Hunt* is still fully equipped to operate at night, but it will be slow going to ensure that we do not hit a floating log. Forty minutes later, we watch the two buoys separate. The divers have reached their 120-foot stop, released the decompression buoy, and started their hang. *Sea Hunt* drifts alongside the buoy in the building current.

By the time the divers surface, we have moved almost a half a mile. It is 25 minutes to sunset. As the three men hand up their stage bottles and the bulky underwater video camera housing, I can sense that they are trying to hide their excitement. Dan Warter says something about a big rock, but I'm not buying it. Finally, they break.

In response to Dan's rock comment, Paul says calmly, but loudly, "No it's not."

The divers let out a torrent of vivid descriptions punctuated by high-fives and fist bumps.

"Did you see the second engine? Did you see the engine stators? They're both split in two at the turbine second stage!" Rob is uncharacteristically enthusiastic in his wordiness.

"How about that first piece of debris, the one sticking halfway up out of the mud. Was that a practice bomb?" Dan Warter asks the other two divers.

"What was that glove doing there? How the hell could it be a glove?" Leave it to Paul, I sigh with a big, goofy grin on my face, to get excited about a glove.

"There's so much stuff—the debris field just keeps going. I think that big, vertical piece is where the cockpit used to be!" Rob yells out as he climbs the last rung of the dive ladder.

They have found the mother-lode of the wreckage. The first recognizable object seen, it turns out, is not a permanent part of the A-6 at all. About halfway between the anchor and the main body of wreckage lies a half-buried practice bomb with its tail sticking out of the mud. The small tail fins are covered in thick, crusty barnacles, making the bomb look twice its original size, yet retaining its telltale shape. Thirty feet from the downline anchor lies a dense debris field, about fifty feet long. It contains 510's two Pratt and Whitney J52 engines, each split in half. Close to the engines are one of the main landing gear and the nose landing gear. The aft equipment bay, the original location of the fiberglass ducting found by Paul, is there too, although we won't know it until I review the video several days later. Many other large parts cannot be immediately identified. There are pieces of wings, but nothing bigger than a person.

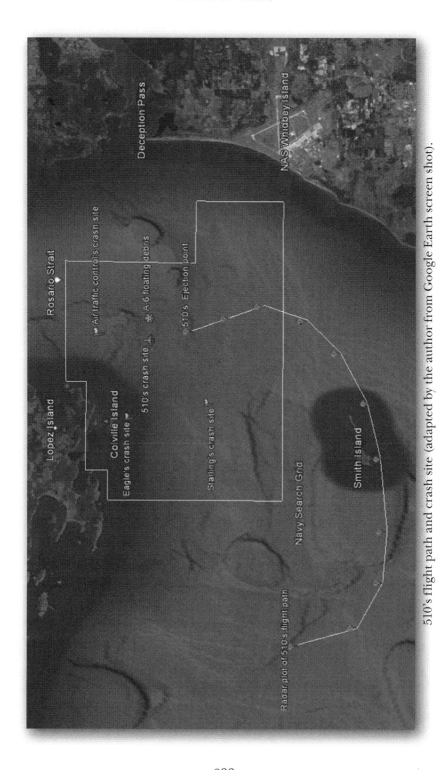

510's flight path and crash site (adapted by the author from Google Earth screen shot).

The Lost Intruder

The lost Intruder might have been relatively intact at water impact, but it has deteriorated to such a degree over the 26-years since then that it's impossible to tell for sure. The debris field is sufficiently dense to indicate that, at one point after crashing, most of the pieces were probably connected. Nothing sits up off the mud any higher than four feet. The timeliness of the find strikes me—how much longer will her crumbling pieces be visible on a recreational sonar unit like the Dragonfly? This might have been my last opportunity to find the lost Intruder, but for a different, wholly unconsidered reason. In a few years, 510, as a practical matter, might cease to exist.

Maritime Documentation Society divers Paul Hangartner, Rob Wilson, and Dan Warter after the October 17, 2015, lost Intruder dive (from the author's collection).

All aboard *Sea Hunt* wear the same ear-to-ear grin. The divers get out of their gear, and we retrieve the decompression buoy and then the downline. The sun goes below the horizon as we point toward Deception Pass some four miles away.

Everybody is in the cabin celebrating when I notice that Dan Warter is missing. I turn the helm over to Dan Crookes and bolt to the cabin door to look for the missing diver. Dan Warter is kneeling on the deck, out of sight just below *Sea Hunt's* aft window frame, fiddling with a piece of gear on his rebreather. I return to the helm with an overwhelming sense of relief—it is much too late in the game to have a problem now.

MK-76 practice bomb at the edge of debris field (photo courtesy of Dan Warter).

My sense of well-being lasts about a minute. I crane my neck over my shoulder to look behind the boat, to "check my six" o'clock position, as learned in the Navy. Dan Warter is vomiting over the side rail. The seas are dead calm, and I immediately know this is bad. I ask Rob to grab an inflatable life vest and help Dan Warter to put it on correctly. Darkness comes quickly and the thought of a physically impaired Dan accidentally falling over the side is terrifying.

Rob returns to the cabin with distressing news—Dan Warter feels dizzy as well as nauseous. Both are early signs of decompression sickness. *Sea Hunt* surges forward heading east as I push up the throttles. Meanwhile, Paul and Josh go on deck to evaluate Dan's condition. They quickly conclude that Dan Warter is probably experiencing the bends. It doesn't matter that his decompression stops were made correctly; it means nothing that Dan hung an extra ten minutes at ten feet breathing pure oxygen as an added safety buffer. We all realize that this sort of thing can occur without apparent rhyme or reason. Dan Warter is bent, and it makes no difference right now why this is so. We need to get Dan medical attention and into a recompression chamber as soon as possible. His condition is worsening before our eyes.

The sun goes below the horizon, and the temperature immediately drops ten degrees. Paul drapes his jacket over Dan Warter's shoulders as he sits on the aft deck breathing pure oxygen from one of the diver's emergency stage bottles. Paul swaps places with Josh, who keeps an eye on the injured diver while Paul re-enters the cabin to call the Divers Alert Network's hotline. We need to start the process of getting the diving emergency management organization to facilitate helicopter transport from the marina upon our arrival. The Divers Alert Network correctly identifies Virginia Mason in Seattle as the nearest hospital with a recompression chamber. While Paul is talking on his cell phone, I ask Dan Crookes to call 911 and have local first responder units meet us at the marina. Dan Crookes suggests that we pull into the state boat launch just short of the marina to expedite the turnover, and I agree. In the meantime, Dan Warter has stopped vomiting and is now able to lay down on the deck, still breathing 100% oxygen, the most effective first aid for decompression sickness. He is alert and conscious, which encourages us to take heart in the probability that this is a minor hit, at least for now.

Sea Hunt races through Deception Pass in the profound darkness that immediately follows sunset, shrouding everything in a blackened cloak until eyes can adjust. We maintain full speed, leaving a horrendous wake, as we pass close abeam the navigation lights of a motoring sailboat traversing the narrow passage opposite our course. I apologize silently, but we have higher priorities. With eyes still adapting to the dark, we turn the corner of Ben Ure Island, continuing uncomfortably fast toward the shore 200 yards distant. Fire trucks shine their headlights directly at us, blinding my vision of the small docks ahead, forcing me to judge where the three tiny fingers of the state boat ramps are by radar alone. Keeping our speed up until I can't stand it any longer, I finally pull back the throttles, then power up both engines in reverse as we pull abeam the narrow middle dock. A half a dozen first responders grab *Sea Hunt's* rails, just in time as our wake catches up with us, buffeting the boat violently up and down.

Two minutes after pulling the throttles back to idle, Dan Warter walks across the aft deck under his own power, still breathing from the stage bottle of oxygen carried in his left hand. He steps onto the dock and is quickly directed to lie down on a collapsible gurney that has been wheeled to the end of the small floating pier.

The phone calls continue, first to Dan's wife, as we wait ten minutes for the helicopter to arrive. Fire engines circle the parking lot with their headlights, creating an ad hoc landing zone, with Dan presumably waiting in one of the two ambulances on the scene. The helicopter circles once, and then in a loud whoosh of rotor blades, we can see navigation lights veer toward the parking lot from out over the water. The thump-thump becomes deafening as the helicopter slows and lands, the only illumination coming from its own spotlight and the headlamps of the emergency vehicles. It is 7:25 pm, an hour after sunset, and less than ninety minutes since Dan Warter surfaced. He is wheeled toward the helicopter.

We all climb back onto the boat, and *Sea Hunt* weaves slowly between darkened vessels at anchor on the return to the marina. I take my time tying up at the fuel dock, close to the steep ramp leading to shore. We begin offloading dive gear at an even pace. The remaining MDS divers get in their cars to make their way to Virginia Mason while I return the boat to the slip. Once again, Christine has let her presence be known. It strikes me that perhaps the lost Intruder prefers to stay lost. It also crosses my mind that it might be wise to respect her wishes. I go home to wait on word of Dan Warter's condition.

Nineteen

...AND MARVEL AT LIFE AS IT'S LED

Dan Warter arrived at Virginia Mason conscious and alert for his first eight-hour session in the recompression chamber. Despite it being a mild case of decompression sickness, Dan compared the symptoms to that of the severe flu: dizziness, constant nausea, and vomiting for days. It was not until the following week, and after a total of four chamber treatments, that all vertigo left him. The total bill for the 100-yard ambulance ride, the 70-mile helicopter evacuation, the four recompression chamber schedules, and hospital care came out to just over $100,000. The Divers Alert Network insurance paid for it all without argument.

The cause of the residual helium bubbles in Dan's body and his ensuing decompression sickness remained a mystery. What set him apart from the rest of the dive team, making him somehow more susceptible to the bends? He followed the exact same dive profile as Rob and Paul, both who had dived the day before, theoretically making them more vulnerable to the bends due to the small amounts of residual inert gas still within their bodies. Dan did not feel ill before his descent. He did not have an equipment or breathing gas irregularity. His decompression procedure that day was by all accounts flawless. Maybe it was triggered by a vindictive Christine's tempestuous post-violation rant, but more likely it was just one of those things that happen in a cutting-edge sport where all the physiological variables have not yet been identified or quantified. Not having hurt anyone permanently is one of the greatest achievements of the lost Intruder project. This was at least the second time that we came damn close.

The lost Intruder continues to hold tightly onto her secrets. The relatively small pieces of the shattered A-6 are far too corroded to point to what caused the hydraulic failure. Nor will we ever know with certainty why 510's landing gear did not come down. The single most significant thing learned about the lost Intruder is its location on the sea floor. I'm not superstitious. I will probably bring the MDS divers back to the site in the future, although my initial inclination is to leave her be. Enough warning shots have been fired across *Sea Hunt's* bow, sufficient hints of impending disaster to highlight the fine line between superstition and common sense.

510's main landing gear, perhaps the largest intact piece of 510 other than the vertical stabilizer, only sits about two feet up from the muddy bottom (photo courtesy of Dan Warter).

There is still no rock-solid proof that 510 was the same jet that skipped off the water near the Philippines. The wealth of circumstantial evidence is enough for me, though. This explanation offers a consistency to 510's history, especially given no alternative theory to explain her perennial mechanical issues. Again, in the end, it matters only to curiosity, and we will probably never know for sure. The only significant trophy, the lost

Intruder's latitude and longitude coordinates, rests comfortably alongside the warm shadows of my memory.

The U.S. Navy has been informed of the find, but they seem disinterested, willing to allow us to do with the wreck what we will. Even so, I have no intention of bringing up 510's pieces. It would be hard, dangerous work with no point.

Pratt and Whitney J52 engine split in two (photo courtesy of Dan Warter).

My brain's images and remembered sensations of 2014's physical pain and struggle of defiance are fading. The human mind's innate resilience tends to push aside past hardships, buffering those that remain within a protective cocoon, a silk parachute to ease the fall. The brain seems to adjust amazingly well to most intrusions, that is, so long as the mind—what I consider to be the contemplative aspect of the brain— is permitted to recover as well. The lost Intruder's overarching lessons recede more slowly, with an intuitive understanding of the most poignant still firmly grounded within me. The lasting value of the discovery of the lost Intruder lies in personal, perhaps even spiritual, growth. The gains that matter are to be found in deep reflection and new friendships. I'm proud to call the MDS divers friends. With 2014's hubris in my Parkinson's impacted boat handling capabilities but a fading memory, friendship is a comfortable and fitting place for pride to once again take form and set.

The lost Intruder project has answered some questions, however, about the Navy's failed search for 510. Would the Navy have fared better had they referenced the air traffic radar logs for the ejection fix? Perhaps. But if the *Salvor* also went along with air traffic control's incorrect post-ejection conclusion based on the wingman's transponder, they would still be looking for 510 over a mile and a half too far to the north. The fact remains, however, that despite the Navy's incorrect ejection fix, they conducted a 29.3 square nautical mile search for 510 with multiple ships using state of the art underwater location gear. The lost Intruder lies well within the six-square-mile high priority sector of this larger search grid. Yet the Navy found nothing.

So, what happened? In my opinion, the Navy woefully underestimated the difficulties in effectively searching local areas of high tidal current. Ben and I did close to the same thing, at least initially, during our 2014 search. But the Navy might have encountered a complicating factor. The Navy grid system was laid out on cardinal headings, possibly limiting search runs to directly north and south, or east and west. This is pure conjecture on my part, but it is a reasonable guess. The prevailing current at 510's actual crash site splits the difference between cardinal headings, running along a southwest to northeast line.

During the summer of 2015, I discovered that the only consistently productive DownVision runs were made with *Sea Hunt* pointing directly into the current, even on days of low tidal flow. Maneuvering into the current allowed for a slow and relatively steady sonar run. By centering the search area based on the direction of the currents instead of arbitrary cardinal headings, *Sea Hunt* was almost always able to make at least half of the runs highly stabilized, allowing the DownVision to depict useful data. In fact, during much of the five days of my successful final survey, only half the sonar runs—those directly into the current—turned out to be useful. The downstream sequences were used to get back into position for the next, more controlled, run into the current. Any data obtained while not going directly into the current was suspect.

The Lost Intruder

If the Navy sonar operator turned his attention away from the task at hand for two seconds, then he would likely have missed 510's wreckage entirely. If the search ship was running with the current, then there would be even less time. The sonar returns are that small. It took me nearly a month during the best weather of the year to carve out five days of flat seas and mild currents suitable for searching a tiny portion of the Navy grid's 29.3 square nautical miles. It was only because of the lessons learned from eighteen months dealing with the Rosario Straits currents that I could narrow the search zone. The key to finding the lost Intruder turned out to be local tidal knowledge, experience, and patience.

Aft half of Pratt and Whitney J52 engine (photo courtesy of Dan Warter).

Five-ten's actual crash site is about two nautical miles from both Lopez and Colville Islands, nowhere near any of the three aircrew distance estimates (Denby Starling's comes closest, but he specified the crash site as being between Lopez and Smith Islands, which it is not). In retrospect, this was one key piece of evidence that the *Salvor* got exactly right: disregarding eyewitness accounts. Five-ten rests more than a half a mile from the closest point that Ben and I searched in 2014.

There might have been another reason why Chris Eagle's "¼ to ½ nautical mile south of Lopez" estimate was so far off the mark. If the point where Chris Eagle's parachute inflated is aligned with 510's actual water impact site and then extended out several miles, the line runs directly through a spot called Davidson Rock. Davidson Rock turns the corner at the southeastern tip of Lopez Island, nearly exposing a shallow reef that literally boils from the strong currents traveling up its slope. Davidson Rock is almost exactly ½ nautical mile south of the very southeastern tip of Lopez Island which is called, confusingly enough, "Point Colville."

Might the bubbling water witnessed by Chris Eagle while swinging in his parachute have emanated from Davidson Rock? Once safely ashore, it would make sense that Chris Eagle would reference a nautical chart to try and pinpoint 510's water impact point. Twenty-five-year-old memories of using "Point Colville" as a reference could have been easily confused with "Colville Island," which lies just a quarter mile to the west of Davidson Rock.

Today, Davidson Rock is marked by a green navigation buoy, but in 1989 it was identified by a concrete block structure on pylons. I know this because it was a source of interesting conversation after a winter storm knocked over the concrete platform sometime in the late 1990s or early 2000s. In the heat of the moment, could the old concrete structure, surrounded by turbulent water, have been mistaken for the fading image of a bubbling A-6 sinking? It would reconcile the precision of Chris Eagle's statement with a logical theory. Could Chris Eagle's distance estimate have been accurate, only to the wrong object in the water? I believe so.

The actual bearing from where ejection occurred to 510's debris field is 347-degrees true north. The distance from the ejection point to 510's water impact is just a shade under one nautical mile; the Navy had that part right.

The cause of 510's total hydraulic failure was never definitively determined, but a theoretical simulation of the accident was formulated on some engineer's desk, probably at the Grumman Corporation. The Backup Hydraulic system was designed to automatically run if either of the two main systems, the Flight or Combined, failed. The three hydraulic systems were engineered to operate independently from each other with one exception: the Backup system replenished its hydraulic fluid from the Combined system.

When the Flight Hydraulic system failed, the Combined system took over the hydraulic workload of the flight controls, and simultaneously the Backup system automatically activated. The engineers hypothesized that if the Backup Hydraulic system pump continued to work correctly, but developed a hydraulic leak, then it would pump overboard not just the Backup system's fluid, but that of the Combined system as well. If this happened, the aircraft's remaining operating hydraulic pumps would be depleted of all hydraulic fluid, possibly in a matter of minutes.

A check valve was installed to isolate the Backup Hydraulic pump from the Combined Hydraulic system as a precaution. The anecdotal evidence indicates that this solved the problem: there was never another instance of a complete hydraulic failure in an A-6. One mystery was left unresolved. Other than the Tabones Rock incident, no reasonable theory was postulated for the failure of the backup landing gear blowdown system.

Many things had to go right that deserve credit for the find. If Ben Griner had not volunteered hundreds of hours to the project, I would not have gained even the basic understanding of sonar operation that I possessed in 2015. Nor would we have been able to discount about ten square miles of coastal waters from the larger possible search area. Without Tugg Thomson's suggestion to request the JAG Investigation, the project would have been dead in the water. Throughout 2014 and 2015, the MDS divers—Rob Wilson, Paul Hangartner, and Dan Warter—were unwavering in their commitment and willingness to risk their lives to dive any contact that appeared even marginally worthwhile. The final DownVision contact was the definition of marginal. Still, I had a gut reaction to the site that Rob Wilson, and perhaps Paul Hangartner as well, soon shared. And, of course, if Paul had not found the key piece of evidence—the fiberglass ducting—we might not have had the confidence in the site to conduct the two critical dives to follow.

An engine cowling stuck on end in the mud (upper left) is one of the tallest pieces of wreckage. A section of canopy glass is in the lower center of the photo (photo courtesy of Dan Warter).

That the MDS divers were willing and able to take on the identification phase, assuming we ever got that far, inspired me to keep parsing at the research. It was only the tenacity of these explorers that kept me looking, both at my desk and in the straits. There are many uses of the word "I" in this account. This is because much of the story is tied to my personal feelings and physical condition. But as far as the actual discovery of the missing A-6 goes, just about every "I" should be replaced with a "we."

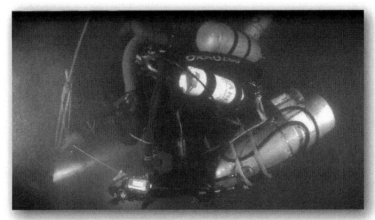

Paul Hangartner preparing to ascend after a dive on the lost Intruder (photo courtesy of Dan Warter).

The Lost Intruder

As complicated as this story is to explain, the process of finding the lost Intruder is still far more easily described than my internal transformations. I know what made me happy before DBS surgery, but that does not mean that I now honestly or wholly understand the fundamental root source of my contented awareness. What does remain powerfully undiluted is a confidence that, somehow, I will get back to that state of being.

I do know, however, that the lost Intruder proved to me that despite physical deterioration, I still mattered, at least in my own mind. The project did not offer false hope or self-pitying dreams in the face of debilitating illness. Instead, the search provided a degree of control over my life. The lost Intruder—as I prefer—or Christine, or 510, still retains a tight fist around her secrets, giving away nothing for free. I only earned an illusion of control step after every painful step of the way. I'm confident that had Christine wanted—if she could "want"—that she would have wanted to remain lost. I also know that without the journey, I would have remained lost as well.

Ultimately, whether called the lost Intruder or 510 or Christine, all mankind's trophies are statues of faded meaning, with their only permanence emanating from the lingering associations that reside in our soul. Perhaps this is the real source of any ghost. Today, many months after the find, the lost Intruder is not obscured by a timeless memory of a young aviator, or of a long ago confident and accomplished diver. There is no great achievement of historical or aviation value, nothing of lasting substance to show for the effort—except, perhaps, Paul's piece of fiberglass ducting. What I do feel close is the victory of the raw battle within me, a sense of focus and the potential to conjure up absolute confidence when needed. I feel engaged and aware. I know that when Parkinson's urges me to quit a task in frustration, I can call out the lying son-of-a-bitch disease and say, "Fuck you—I can do this." But only to a point.

The lost Intruder project was more than a salve to an aching spirit, although it accomplished a little of that. It enabled me to better understand who I am, even as that definition changes before and within my eyes. But all things change. When the disease's progression puts my failing body back to where it was pre-surgery, I will no longer be the same person as during the first battle; I will be different in unpredictable ways. Taking a lesson from the two opposite sides of my nemesis, my "Janus"—the twin faces of Parkinson's disease and Christine—it's evident that a moving target is always the hardest to hit. This is as applicable now as when an impossibly younger me flew A-6 attack bombers.

By finding the lost Intruder, I beat my internal adversary, allowing me a fighting chance at a meaningful life. Christine had become the disease. To understand her—for me to become aware—the lost Intruder needed to be found.

This has taught me that life's truly important answers are already within me. The search for the lost Intruder was at best a surrogate, a catalyst to help coax out the clues to my inner exploration. Much like my failing body, the lost Intruder suffered from an ailment brought on years earlier and undiagnosed. As a practical matter, I do not believe it is an exaggeration to say that the project saved me from a premature death. It kept me alive by enabling me to circumvent Parkinson's harshest blow, depression, both before and after DBS surgery when I was most vulnerable. But, as important as this is, it alone does not do the lost Intruder project justice.

The personal challenges encountered forced me to honestly study and take ownership of my identity. It allowed me to accept who I was beneath my deteriorating physical façade. By losing an irrational fear of failure, and by picking a challenge without the cluttering influence of material value, I inadvertently opened a door leading to a liberating higher consciousness. Perhaps someday, as I continue down this path, it may result in actual enlightenment. The experience showed me how to be happy, a state of being not experienced in little pieces, but through a comfortably durable sensation month after month that no amount of pain could mask.

Happiness is elusive and not necessarily found in the continuum of circumstance through a shared formula. But I do remember, and it is much easier to find the missing piece of a puzzle when one knows that it exists. I'm not certain if such a struggle is necessary to force this discovery for everyone, but I am sure that it was for me. How quickly happiness slipped away after surgery, once physically more comfortable and relatively pain-free, supports this belief. When I re-negotiate the path to happiness again, it will be because of the search for the lost Intruder, not despite the experience. Fighting the damn fog every dragging step of the way helped me discover myself honestly, and, in the process, it taught me real honesty.

Effort is measured in inches and feet and miles, but success is only determined by the will to keep trying. Without realizing it, I had bypassed the material goals that dominate our society and instead looked for reward from within. And, for a time, I found that prize through a sense of contented acceptance and awareness: it was a profound spiritual peace that led to my happiness.

The practical—those who discount the spiritual aspect of the journey might call it the real world—value of this experience can be found in a single word: continuity. My role in the project, as a researcher, boat operator, and leader were all new, although familiar in that they were intrinsically linked to my passions of old, diving and flying. More than a distraction, the quest for the lost Intruder tied together these past lives in a continuous stream of consciousness, allowing me to maintain a gut-level instinct of who I was—and am—regardless of Parkinson's or any disease. I had flown that jet from the U.S.S. *Ranger* as a young man; I had made dives similar in complexity and hazard to that of the MDS volunteers before my diagnosis. These events, as detached as they are now in both years and my physical ability, actually happened.

From my accelerating Parkinson's symptoms of 2014, to DBS surgery, to the post-surgical symptom-free honeymoon and the rapid physical changes of six weeks until battery turn on, the lost Intruder provided a critical stability. During the year to follow, the project provided a sense of continuity with pre-surgery Parkinson's challenges, as well as with life long before my hand first began to shake. It allowed me to maintain a tenuous control over mind and body while being wrenched through a series of desperately ratcheting changes. The search provided for a glimmer of future brightness without neglecting the present. It didn't just make me feel sane; it genuinely kept me there as well.

Everybody has an overarching identity of some sort. Skin deep identification with a career, skill, religion or a group of any kind is often desperately sought, maybe in part to quell our deep insecurities as humans in a complex world. Most identities are not consciously considered, never mind well-thought out.

Moving from a vague grasp of identity to a purposeful self-awareness—a denial of self, even, a questioning of physical existence—does not just happen: it isn't stumbled upon in some accident of fate. It should be an inner conquest, a transformational journey. The old identity can't be just disregarded or bypassed. It needs to be exposed to the light and decimated, torn from the soul in a surrender of ego. The new cannot be built on top of the old but should be permitted to carve out its own niche in the psyche's landscape and flourish unfettered, with minimal influence from the past. It should be a celebration of impermanence and the changing nature of the world.

During the search for the lost Intruder, I did not in any central way consider myself first an explorer or—for that matter—a writer, father, or husband. Or a fading diver or former pilot. And I sure as hell didn't consider myself first and foremost a Parkinson's victim. But my playful demon came far closer to winning than I realized at the time. Instead, my predominant identity had become the guy who was fighting Parkinson's disease. Not a subtle difference, but a chasm of meaning that was nonetheless bridged by that single word—Parkinson's.

This realization only came more than a year after DBS battery turn-on, and like many epiphanies, it was shamefully brilliant in its conspicuousness upon recognition, which explained a lot about my post-surgery struggle. I still had Parkinson's, and it would get worse. There was no permanence here, which, of course, highlights the fact that there is no permanence anywhere.

What I was desperately missing—what I had been clinging to—was the identity lost due to successful surgery, which gives me the opportunity to create my new self. I only realized how much I missed the simplicity of being the guy who was fighting Parkinson's disease after completing this book. The process of writing was exactly what was needed to put it all in perspective. It became apparent that the lost Intruder got me through a hell of a lot more than the surgery. It offered a new lease on a life worth living, but only if I grabbed hold and began to fully explore my inner self. This is where all true answers lie.

And that is what I'm trying to do. So far, it appears to be working.

Afterword

It would not be for more than a year before *Sea Hunt* successfully returned divers to 510's wreckage. Inclement weather contributed to the delay, as did life's ordinary commitments. Death played a part as well. The passing of former MDS leader, Ron Akeson, a highly respected and well-liked man, shook the Society, producing long term effects that still linger today. Ron had been the second MDS member from a cadre of less than a dozen to meet with disaster. The first—Mark Thuene, an attorney, and avid underwater explorer—died while on a deep Great Lakes wreck dive several years before my search began. The loss of two close friends was enough for Josh Smith; he quietly stopped diving and sold his equipment.

These tragedies set a somber stage. When a third MDS diver, a man who had played no part in the lost Intruder project, was killed in a 2016 diving accident, Dan Warter decided to take a break from diving, perhaps permanently. There is no doubt that Dan's brush with decompression sickness helped with his decision as well. The explorations of Rob Wilson and Paul Hangartner seemed unaffected by the calamitous events, but nothing in life is as it appears, as I was soon to be reminded.

Fate continues to intervene in all our lives. It was only after *Sea Hunt's* return to the lost Intruder on January 24, 2017, that I was told some startling news. Paul Hangartner had just finished removing his gear after the dive. He looked haggard to me, but he did work nights, which often required that he join the daytime technical dive schedule with little sleep. There was something familiar in his eyes, though, a deep-seated tiredness that I remembered well from looking in the mirror pre-surgery. I asked Paul if he was alright. His reply jolted me from life's distractions.

"No, I'm not alright," he said. Paul had been diagnosed a year and a half earlier with widely metastasized Neuroendocrine Cancer. If he is fortunate enough to be responsive to the only recommended treatment, which merely delays death marginally, Paul had two to three years left to live.

Paul's dive that day had been wildly successful. Despite the winter calendar, we had perfect weather and current, allowing for a precision drop of the downline directly to the spot where the Intruder's engines had been discovered. With Dan Warter no longer available to video the wreckage, Rob had arranged for another experienced technical diver, John Sanders, to join the dive team. John surfaced with video of the Intruder's empennage, less than two dozen feet across the muddy bottom from the A-6 engines. The port side horizontal stabilizer was intact and attached to the remnants of the vertical stabilizer, which despite lots of deterioration, was still recognizable.

The location of 510's broken-off empennage so close to the rest of the wreckage seems to indicate that tail separation occurred at surface impact, allowing the section to drop rapidly to the bottom. The tail also turned out to be the spot where my exploratory grapnel hook eventually caught back in the summer of 2015. John Sander's video shows Paul turning the corner of the tail section, just before letting out an audible peel of laughter when coming face-to-face with the grapnel hook. Paul had been joking about spending valuable minutes of dive time looking for the worthless anchor just moments before entering the water.

The news of Paul's impending demise left me speechless. As I retraced the timeline, it struck me that he probably learned of his incurable disease about the same time we found 510. He must have told Rob, but no one else. When Paul said goodbye to me after the dive, I gave him a big, rough hug, just as he had done for me before my surgery. I told Paul that he was inspiring, still attacking life even as death approached. His reply surprised and moved me:

"Peter, you know what it's like." He looked directly at me, eyes wide with comprehension. "What am I going to do, curl up in a ball and die? You didn't."

His words were unambiguously those of a contender.

No, I didn't and he won't. Paul wants to dive the lost Intruder at least once more while still able. I've promised myself to do whatever it takes to make sure that he gets his wish. It's the least I can do for an old shipmate, a truth-teller who only requests to be allowed to follow his own path to inner peace. None of us need look very hard for inspiration in our lives; it surrounds.

The Lost Intruder

In the struggle for direction in our lives, as we attempt to "fly the plane," it is important to realize that ultimately there is no control. Let go the illusory power of a command over fate while breathing fully and deeply, accepting unfolding events on their own terms. Marvel at life as it's led, soaking in the wonders of a simple breath.

Cheers,
Peter Hunt
May 2017
Whidbey Island

A note on the source of chapter titles.

All chapter titles were taken from the below poem, written during the height of my discomfort. It is not a sad poem, but to understand its optimism, it should be read thoroughly. The poem describes the sensations and feelings during a severe transition from dyskinesia to dystonia and back—and my fight against the "eager specter's" multiple daily visits.

<u>A Bump in the Road</u>

Dawn's patiently eager specter,
Schemes sickly in brightening day,
Cruel with well-honed diversion,
Silently awaiting its prey,

Decomposed jabs of distraction,
Blunt needles plunge slowly and deep,
Masking a stranger's true nature,
A shadow's faint outline of sleep,

A fluttering string of confusion,
Bind in contortions of sin,
Twisting in circled indifference,
Preparing for conquest within,

In painfully familiar direction,
With tactics pitifully weak,
The turn of temptation approaches,
The seductress too guilty to speak,

No duty or pity or honor,
Obscure misdirection or doubt,
Predictable rhythmic confusion,
A pathetic attempt at a rout,

Peter M. Hunt

Eyes locked in mortal-gripped vision,
Transfixed beyond images past,
The black-soul tempest approaches,
Stand firm aside tiller and mast,

Waves beat in graceless harsh meter,
Caulked nerve holds muscle to steel,
Transparent enticement no match,
For unshakeable ken of the real,

Will free of purposeful measure,
With instinct manning the rail,
Lungs thirst for the round to be over,
For the fighting-mad core to prevail,

In timeless cacophony's wisdom,
The test runs its course and subsides,
Sudden end to intensity's shudder,
Quell urgent desires to die,

The battle won with the knowledge,
That unwinnable war lies ahead,
With ease forget the last hour,
And marvel at life as it's led,

It's just a few portions in hours,
Spread even throughout the day,
Not worthy of long term attention,
A bump in the road on the way.

by Peter Hunt
June 2014

The author at *Sea Hunt's* helm over the wreckage of the lost Intruder on January 24, 2017 (from the author's collection).

Peter Hunt was born in New York and spent six years of his childhood in Athens, Greece where he started diving in 1978. He graduated with a history degree from Brown University in 1985 before joining the Navy and training as an A-6 Intruder attack pilot. Hunt completed three aircraft carrier deployments to the Persian Gulf, Indian Ocean, and Western Pacific during ten years of active duty. His military awards include the Distinguished Flying Cross and three Air Medals for combat action during Operation Desert Storm. After leaving the Navy, Hunt worked as a United Airlines pilot until being diagnosed with Parkinson's Disease in 2005 at age 43. The father of two adult children, Hunt holds a Masters in Strategic Planning for Critical Infrastructure from the University of Washington and lives with his wife on Whidbey Island, Washington. He is the author of *Angles of Attack, an A-6 Intruder Pilot's War*, and *Setting the Hook, a Diver's Return to the Andrea Doria*. For more information, please go to www.peterhuntbooks.com.